AF725093

Science Communication on the Internet

Pragmatics & Beyond New Series (P&BNS)

ISSN 0922-842X

Pragmatics & Beyond New Series is a continuation of Pragmatics & Beyond and its Companion Series. The New Series offers a selection of high quality work covering the full richness of Pragmatics as an interdisciplinary field, within language sciences.

For an overview of all books published in this series, please see *benjamins.com/catalog/pbns*

Editor

Anita Fetzer
University of Augsburg

Associate Editor

Andreas H. Jucker
University of Zurich

Founding Editors

Jacob L. Mey
University of Southern Denmark

Herman Parret
Belgian National Science Foundation, Universities of Louvain and Antwerp

Jef Verschueren
Belgian National Science Foundation, University of Antwerp

Editorial Board

Robyn Carston
University College London

Thorstein Fretheim
University of Trondheim

John C. Heritage
University of California at Los Angeles

Susan C. Herring
Indiana University

Masako K. Hiraga
St. Paul's (Rikkyo) University

Sachiko Ide
Japan Women's University

Kuniyoshi Kataoka
Aichi University

Miriam A. Locher
Universität Basel

Sophia S.A. Marmaridou
University of Athens

Srikant Sarangi
Aalborg University

Marina Sbisà
University of Trieste

Paul Osamu Takahara
Kobe City University of Foreign Studies

Sandra A. Thompson
University of California at Santa Barbara

Teun A. van Dijk
Universitat Pompeu Fabra, Barcelona

Chaoqun Xie
Fujian Normal University

Yunxia Zhu
The University of Queensland

Volume 308

Science Communication on the Internet. Old genres meet new genres
Edited by María-José Luzón and Carmen Pérez-Llantada

Science Communication on the Internet

Old genres meet new genres

Edited by

María-José Luzón
Carmen Pérez-Llantada
University of Zaragoza

John Benjamins Publishing Company
Amsterdam / Philadelphia

The paper used in this publication meets the minimum requirements of the American National Standard for Information Sciences – Permanence of Paper for Printed Library Materials, ANSI Z39.48-1984.

DOI 10.1075/pbns.308

Cataloging-in-Publication Data available from Library of Congress: LCCN 2019036906

ISBN 978 90 272 0466 0 (HB)
ISBN 978 90 272 6179 3 (E-BOOK)

© 2019 – John Benjamins B.V.
No part of this book may be reproduced in any form, by print, photoprint, microfilm, or any other means, without written permission from the publisher.

John Benjamins Publishing Company · www.benjamins.com

Table of contents

CHAPTER 1

Connecting traditional and new genres

Trends and emerging themes

María-José Luzón and Carmen Pérez-Llantada
University of Zaragoza

Genre theory is an all-encompassing explanatory framework for understanding texts and the contexts and social structures they are embedded in and, as such, it has a strong pragmatic orientation. Genre-based communication has been analyzed from a wide range of perspectives and approaches. According to the social rhetorical view of genres, the approach taken up in the New Rhetoric School, genres have been defined as "recurring rhetorical situations" (Miller 1984: 152) and as frames for social action that shape "the thoughts we form and the communications by which we interact" (Bazerman 1997: 19). On the other hand, the perspective of English for Academic and Professional Purposes and Languages for Professional Purposes characterizes genres in terms of regularities in communicative purposes, functional moves and linguistic features – see the seminal work by Swales (1990) and Bhatia (1993). In further theoretical developments these authors propose the view of genres as mechanisms that give meaning and function to particular texts with intended purposes and recipients (Bhatia 2004; Swales 2004). Research in the domain of professional communication likewise highlights that rather than "text types," genres are "ways of recognizing, responding to, acting meaningful and consequentially within, and helping to reproduce recurrent situations" (Bawarshi and Reiff 2010: 3).

"Genre" has also been a key term used in empirical research to understand issues such as register, genre-conventions and style in scientific and research writing (Paltridge 1995; Biber and Conrad 2009). More recently, digital genres have become an incipient focus of critical enquiry, as evidenced by sound empirically-based corpus-driven theorizations of specialized registers in the age of "mass literacy," as Biber and Gray put it (2012: 314). Considering the aim and scope of the present volume, it is also worth noting here that genre theory has also been applied to the analysis of computer-mediated communication (CMC). Studies on digital genres have underscored the need to revisit and expand the concept of genre to incorporate media elements and accommodate new online forms that

https://doi.org/10.1075/pbns.308.01luz
© 2019 John Benjamins Publishing Company

defy pre-digital conceptualizations (Askehave and Nielsen 2005; Bateman 2008; Miller and Shepherd 2009; Heyd 2016).

Genre has also been defined as a powerful analytical tool for theorizing scientific discourse and an essential concept to understand how members of specific discourse communities use language to accomplish social actions and respond to recurring rhetorical situations in a particular context. Rather than being static forms, genres are "inherently dynamic rhetorical structures that can be manipulated according to the conditions of use" (Berkenkotter and Huckin 1995: 6). Existing genres evolve and new genres emerge to address new social exigences and to adapt to changes in community membership, audiences, disciplinary activities and methodologies, media and technology, disciplinary values or public attitudes to science. In the digital era, scholars are progressively using online genres to respond to new rhetorical exigences and meet the increasing demands from funding agencies for openness of the research process and research outcomes.

The emergence of Web 2.0 technologies has heralded a new era of research activity and science communication, characterized by the integration of these technologies in the knowledge production and communicative practices of disciplinary communities. The specific technological affordances of new media are revolutionizing the ways in which researchers produce, represent, re-use and share information and knowledge. ICTs and the Web have provided a plethora of new platforms, tools and forms of communication for academics to conduct research, make research objects (e.g. software code, data sets) available and reusable, discuss and share scholarly work, both with peers and the interested public, engage citizens in the research process, and increase the visibility and reliability of their research. Some examples of these forms are online papers and enhanced publications (Breure et al. 2011; Aalbersberg et al. 2012; Pérez-Llantada 2013), online pre-prints (Delfanti 2015), wikis (Kelly and Miller 2016), blogs (Luzón 2013, 2017; Mahrt and Puschmann 2014), Twitter (Smith 2015; Yeo et al. 2017), research group websites (Luzón 2018), academic homepages (Hyland 2011), academic social network sites (such as ResearchGate or Academia), online videos (Erviti and Stengler 2016), open notebooks (Carter-Thomas and Rowley-Jolivet 2016; Wickman 2016), webinars (Ruiz-Madrid and Fortanet-Gómez 2017), or discussion forums.

This ever increasing number of online genres for science communication, which result from the possibility to harness the affordances of the Internet to respond to new rhetorical exigences, promotes new communication and disciplinary practices. These changes should be understood within the wide socio-political context of current research endeavour and in relation to assumptions about the role of the scientist today and about the importance of science communication to gain the public's support for science and scientific research. The explosion of digital genres is inextricably linked to the current societal and political push for the democratization of science and the sharing of research. Open Science, "the idea that scientific knowledge of all kinds should be openly shared as early as is practical in the discovery process" (Nielsen 2011), is now promoted (or required) by many funding agencies and research institutions, as it contributes to reproducibility and accountability of research, dissemination of knowledge, re-use of scientific data and better return on investment of public funds (Ball 2016). Open Science involves giving other researchers free access to the research process and the products of research (research publication and scientific data, source materials) under the assumption that knowledge is not a commodity but a public good and that open inquiry, facilitated by the interoperability of freely accessible science literature and scientific data, is "at the heart of scientific enterprise" (Boulton 2012). Open Science is also concerned with making research results accessible and comprehensible for a wider audience, which motivates researchers to engage in new forms of communication that enable them to explain their research to the public and gain their support.

Digital genres facilitate Open Science participatory practices and offer researchers new opportunities for networking, open discussion and interaction, and the creation of connections with like-minded peers and interested publics. The Internet provides spaces and tools for the public dissemination of science, for making a given piece of research available to a diversified audience, by propagating it (in various forms) through several platforms, and for engaging citizens in scientific research. These tools also account for several changes in peer-review practices, e.g. open review reports, open review by the wider community, open discussion between reviewers and authors (Ross-Hellauer 2017; see also Breeze this volume). Although most quality research is still published in high-impact pay-walled journals, which contributes to the limited access to research output by researchers in the global South and thus to knowledge divide, the increasing adoption of digital genres to share and discuss research may help to challenge this status quo.

As a consequence of the high impact of new technologies on science communication, in the past few years there has been an increasing interest in the genres used by researchers to communicate science online, as attested by the number of recent conferences and seminars focusing on how the affordances of the Internet are changing genres. If we look at the wide range of international academic events taking place, we see that the impact of technology on research communication has clearly become a major focus of attention. The topics treated, e.g. issues of knowledge dissemination, hybrid digital genres, the communicative immediacy of the medium and the spectrum of genres on the web, have gained and are gaining momentum. Incipient interest in technology mediated genres in the scientific domain can also be found in the edited volume *Science and the Internet* (Gross and Buehl 2016) and in relevant journals such as *English for Specific Purposes*, *Computers and Communication*, *Journal of English for Academic Purposes*, *Language Learning and Technology*, *ESP Today*, *Written Communication*, and *Discourse, Context & Media*, to name a few.

Despite the growing research on academic digital communication, most studies have focused on individual genres, paying less attention to their place in the ecology of genres used by researchers and research units (research groups, research centers) to mediate their research activities, or to the role of online genres on the communication agenda of these researchers. However, as scholarly communication practices are being dramatically transformed by networked communication, it is becoming even more evident that digital genres are not autonomous or standalone entities and that they are related to other (traditional and digital) genres in various and complex ways.

Genre studies have shown that genres interact with other texts to achieve social goals and work together to mediate a given activity (Bazerman 1994; Spinuzzi 2004). Genre analysts have proposed different models of genre relations, some of them closely related: sets and systems (Bazerman 1994; Devitt 2004), hierarchies, chains and networks (Swales 2004), repertoires (Orlikowski and Yates 1994; Devitt 2004), ecologies (Spinuzzi 2004) and ecosystem (Casper 2016). While some terms, such as "repertoire," i.e. the set of genres used by a discourse community to achieve all its purposes (Orlikowski and Yates 1994; Devitt 2004), are useful to get a comprehensive understanding of the shared communicative practices of a community, other terms are more restricted and refer to relations between a limited set of genres used by the community. Terms such as "genre system," "genre set" or "genre chain" are used to refer to genres that are often engaged in by members of a discourse community in a sequential order, and where one genre depends on the use of others. For instance, Devitt (2004: 56) defines genres system as the "set of genres interacting to achieve an overarching function within an activity system."

The digital genres used by members of particular communities are part of their repertoire of genres and interact with and co-evolve with other traditional genres. Some of them are the result of importing print genres into the online medium and therefore are related to their print predecessor by a process of remediation (Bolter and Grusin 1999). In addition, the affordances of the web (hyperlinking, heightened multimodality, global reach, interactivity, ease of reusing text) and the network structure facilitate interconnections between genres. In the web a single genre may embed and link to multiple online genres, creating contingent relations and shifting networks of interacting genres to accomplish specific goals. The affordances of digital media transform the space in which genres exist and create close relations between hyperlinked genres (e.g. a scholar's home page and his/her online research papers) and between "neighbouring genres" (Heyd 2016: 96), which coexist in the same space, e.g. a Wikipedia article and Wikipedia talk pages. Sets of different multimedia genres can be "linked together in locally situated ways," forming what Prior (2009) calls "mediated multimodal genre systems." Genres are also linked to other genres to which they respond and on which they comment, which facilitates immediate access to the other text.

Since genres in digital environments interact with other genres in manifold and complex ways, research on digital scientific discourse should move from the analysis of single genres to the study of the synergies among digital genres and of the relationships between digital and traditional genres. In this context, this volume seeks to contribute to filling the gap in publications on science communication online, by focusing on one essential aspect to understand genres in general and digital genres in particular: their relations with other genres. In the following sections of this introductory chapter we briefly present the current state of research on digital genres for science communication and on their interaction with other genres. We will provide an overview of research on remediation of print genres in the digital media, on the hybrid nature of many digital genres, and on the blurring of boundaries between the scientific community and the public in digital genres. In the final section an overview of the chapters in this volume will be provided.

1. Remediation of print genres

Many digital genres result from the remediation of print genres, i.e. existing genres are imported into new media or they evolve into variants enabled by the affordances of the new medium (Bolter and Grusin 1999; Crowston and Williams 2000; Bawarshi and Reiff 2010). By this process, genres are relocated, transformed, combined, and repurposed to fit in a new context.

An interesting example of genre remediation is the open science notebook (Carter-Thomas and Rowley-Jolivet 2016; Wickman 2016). Carter-Thomas and Rowley-Jolivet (2016) compared a monologic (i.e. readers could not post comments) online notebook with the paper notebook traditionally used in labs. The comparison reveals that the digitality of the online notebook has an impact on its linguistic features: language is more informal and emotive than in the paper notebook and presentational features are strategically used to create a sense of immediacy (e.g. results are presented as "breaking news"). The open science notebooks in OpenWetWare analyzed by Wickman (2016: 12) are tools for the open sharing and discussion of information through digital media. In addition to serving to record laboratory practices and circulating and posting information online for public use, they can be modified by users other than the original producer. That is, they function as "as a dynamic rhetorical space within which groups of scientists and other stakeholders can present, discuss, revise, and circulate information." Since these open notebooks accommodate wider audiences and respond to novel exigences related to open access and data sharing, Wickman (2016: 17) raises the questions whether they "actually constitute an extension of the genre or, more radically, a complete break from it."

Other academic genres remediate traditional genres by exploiting the multimodal affordances of the Internet, e.g. video abstracts (Spicer 2014; Plastina 2017), online lectures (Bernad-Mechó 2015) or webinars (Ruiz-Madrid and Fortanet-Gómez 2017). Video abstracts are defined in terms of their affordances: they are short videos in which authors "briefly communicate their research through a more personal, media rich medium that is better adapted for Internet sharing" (Spicer 2014: 3), thus improving their visibility and accessibility (Plastina 2017). They supplement written abstracts on the journal websites, but they may also be hosted on *YouTube* in order to reach a wider audience. In her study of a corpus of written and video abstracts from online medical journals, Plastina (2017) investigated how the rhetorical structure of the written abstract in transformed in the video abstract and how the written abstract is re-articulated multimodally in the video abstract. Interestingly, she also found that the video abstract is a hybrid genre which incorporates elements not only of the written abstract, but also of other genres like the conference presentation or the interview. The role of multimodal affordances has also been explored to define the webinar (Ruiz-Madrid and Fortanet-Gómez 2017) and the online lecture (Bernad-Mechó 2015) and to determine how these online academic genres differ from their non-digital counterparts. Bernad-Mechó (2015) drew on Multimodal Discourse Analysis (MDA) to study the linguistic, paralinguistic and kinesic features present in the metadiscursive elements that lectures use to link the different sessions within online courses. He also examined how lectures are reconfigured on the web by analyzing the structure, organization and interactive elements of the platform where the lectures were hosted.

Finally, the research article, the most important genre for communication within disciplinary communities, has also been reconfigured in several ways when it has migrated to the web. A clear example of harnessing the affordances of the web is what has been called "enhanced publications" (Bardi and Manghi 2014; Harmon this volume), "Rich Internet Publications" (Breure et al. 2011), or "The Article of the Future" (Aalbersberg et al., 2013; Pérez-Llantada 2013). All these terms are used to refer to publications that improve not only the printed article, but also the online article, in several ways. The digital environment makes it possible to enhance the conventional article with additional material (e.g. videos, interactive maps, data sets, animations, links to citations and to other supplementary texts, dynamically changing tables) and tools (social networking sites, spaces for readers' comments, online polls) (Breure et al. 2011; Bardi and Manghi 2014). The term "Rich Internet Publication (RIP)" is used by Breure et al. (2011) to refer to digital publications which diverge so much from conventional publications that should be considered a new genre, rather than an "enhanced publication."

These are publications that share features of information integration (i.e. integrating interactives like slideshows, image galleries or query interfaces), visualization and exploration (i.e. nonlinear reading) with Rich Internet Applications. As for the "Article of the Future," it is a project intended to "offer an optimal platform to communicate science in today's digital world." This can be done by enhancing the online article in three directions: "Presentation – offering an optimal online browsing and reading experience, which is a basic requirement for online reading and for any further enhancements; Content – supporting a richer pallet of author-delivered material, including multimedia files, scientific data, and computer code; Context – connecting the online article to trustworthy scientific resources to present the reader with relevant information in the context of the article" (Aalbersberg et al. 2013: 236–237). As these enhanced publications illustrate, the process of remediation often results in the emergence of multi-genres.

2. Multi-genres, add-on genres, and hyperlinked genres

The web offers platforms where genres already existing in printed media and new genres can co-exist and interact in complex ways in a single space. Many digital genres are multi-genres or hybrid genres, formed from combinations consisting both of already-existing and new genres which act together to mediate the researchers' activity, and usually involving multiple modes and media (Crowston and Kwasnik 2004; Casper 2016; Pérez-Llantada 2016; Luzón 2017; Mirović, Bogdanović and Bulatović this volume).

Traditional genres that have migrated to the web can be improved with new digital genres. The OpenCourseWare lectures, for example, incorporate in a single platform video-recorded lectures, speech transcripts, supporting slides/texts, class syllabi, assignments and quizzes (Crawford Camiciottoli 2018). Another clear example of a traditional genre enhanced with add-on genres is, as has been discussed above, the research article. Various genres and discipline-specific research tools may be added (by embedding or linking) to the online article for different purposes. One of these purposes is to share all the outcomes and data of the research, so that they can be re-used by others (Breure et al. 2011; Aalbersberg et al. 2013; Bardi and Manghi 2014). Enhanced articles may embed or link to supplementary material, data sets, references and links to relevant databases, software to run an experiment (e.g. interactive map software in Earth sciences). Other add-on genres are intended to increase the visibility of the research reported in the paper, promote it, or make it accessible to diverse audiences, e.g. graphical abstracts (Hendges and Florek this volume), video abstracts (Spicer 2014; Plastina 2017), author summaries (Breeze 2016) and audioslides (Yang 2017). All the add-on genres above answer different needs than the traditional abstract: audioslides, for instance, are short, webcast-style presentations which give authors the opportunity to promote their work, summarize their research in their own words, and show the relevance of their findings (Elsevier 2018); author's summaries or lay summaries are synopses written for a wider, non-specialist audience or students (Kuehne and Olden 2015; Breeze 2016). Kuehne and Olden (2015: 3585) highlight the important role of these summaries for outreach by stating that they will "increase the visibility, impact, and transparency of scientific research" and create "direct pathways between scientists and the general public" (Kuehne and Olden 2015: 3585).

Other online multi-genres (i.e. combinations of various genres on a single platform to facilitate uptake) are native to the web, e.g. research group blogs (Luzón 2017) or academic websites, like research group websites (Luzón 2018), European research project websites (Lorés 2018), or the websites of scientific magazines (Mirović, Bogdanović and Bulatović this volume). In her study of research group blogs written by multilingual scholars, Luzón (2017) found that various genres are integrated and coordinated to accomplish complex objectives, among them to publicize the group's research and make it available to the disciplinary community, and to connect with the interested public. Academic websites are also multi-genre platforms, where information in different formats are combined to reach and connect with various audiences (e.g. the international disciplinary community, public audiences) (Luzón 2018). Some of these websites provide a clear example of the erosion of boundaries between experts and the interested public (Trench 2008).

Embedding and hyperlinking also facilitate the interrelatedness between online research articles and genres that provide feedback on the research reported in the article (see Breeze this volume). Casper (2016) analyzed different tools for online feedback (comments, notes, e-letters) and found that they interacted with online research articles in two ways: they change the discursive environment of the research article and the may bring about changes in the text of the research article itself, e.g. notes can be used by readers to question parts of the article or to expand on an argument. Unlike their printed counterparts, these feedback genres are easily accessible to all the readers of the online article, thus changing the conditions of their uptake (Casper 2016). The hyperlinking capabilities of the Internet also make it possible to link an online article and a text evaluating or commenting on it on a different platform. Sidler (2016) and Buehl (2016) examined the role of blogs as platforms for post-publication review of research articles. Blogs have been used as tools for experts to publicly refute an article's claim and as forums for the discussion of published results, with contributions both from experts and lay publics. Hyperlinking enables an immediate and one-click connection between the article and the commenting blogpost, which facilitates public discussion of published work and alters the established peer review system.

3. Genres for public or diversified audiences: Context collapse

The Internet has also blurred the boundaries between the scientific community and the public and between the genres intended for audiences with different degrees of expertise (Trench 2008; Miller and Fahnestock 2013; Smart 2016; Kelly and Miller 2016; see also Maier and Engberg; Mirović, Bogdanović and Bulatović; Orpin; and Reid and Anson this volume). Miller and Fahnestock (2013) consider that a question that needs to be examined in future research in genre analysis is "how science accommodates public audiences, given the increasing access by the public to primary scientific genres through online media, as well as the increasing potential for interaction between scientific and public communities." With their ability to reach wide audiences, digital genres facilitate dissemination both within and beyond traditional specialized audiences thus leading to "context collapse" (Marwick and Boyd 2011), i.e. the flattening of diverse audiences into one, which forces users to meet the expectations of these diverse audiences simultaneously.

Kelly and Miller (2016) provide evidence for this erosion of the boundaries between expert and non-expert spheres of discourse by analyzing what they call "parascientific genres," a notion adapted from Kaplan and Radin (2011), i.e. genres "that function alongside traditional genres of science communication in that they borrow scientific authority and knowledge structures from the realm of science but operate without the gatekeeping and traditional reporting forms of internal science communication" (Kelly and Miller 2016: 221). A good example of parascientific genres are science blogs. This is one of the online genres for science communication to which more attention has been paid (see e.g. Kjellberg 2009; Luzón 2013; Mahrt and Puschmann 2014), with particular focus on their role as open spaces for communicating science to a diverse audience (with different degrees of expertise) (Luzón 2013) and for the development of conversations involving participants within and without the scientific community (Luzón 2013; Bondi 2018). Given the huge potential of digital genres for public outreach, researchers are analyzing how scientific discourse is recontextualized into some of these genres, e.g. science blogs (Luzón 2013), crowdfunding proposals, i.e. proposals seeking to persuade individuals to provide funding for a project (Mehlenbacher 2017).

Another topic of particular interest when considering how scientific knowledge is recontextualized into digital genres is the role of multimodality in connecting with various audiences. Recently, video genres intended to disseminate knowledge have received some attention (Riboni 2018; Rowley-Jolivet and Carter-Thomas this volume). Riboni (2018), for instance, examines YouTube vlogs posted by scholars to identify the discourse strategies that they employ to facilitate comprehension of their research by the general YouTube public. The analysis of the recontextualization strategies used in these highly multimodal genres has also been approached from the perspective of MDA (multimodal discourse analysis) (Luzón 2019; Mirović, Bogdanović and Bulatović this volume). Luzón (2019) analyzes how multimodal strategies are employed in videos intended to recontextualize disciplinary knowledge for an interested public and how the affordances of different modes contribute to accomplishing knowledge dissemination.

4. Chapters overview

In this section we present a brief overview of the contributions to this volume. The different chapters explore digital genres for science exchange and dissemination focusing on their relation with other genres used by disciplinary communities and their function in the existing repertoires of genres of these communities. Theoretically, this volume brings together research on genre from different traditions (English for Specific Purposes and Rhetorical Genre Studies) and by authors who have explored genres from different perspectives. Methodologically, considering that the world of genres is becoming increasingly complex, the research on genres on the Internet compiled in this volume moves beyond these broad theoretical perspectives and builds upon the complementarity of analytical tools and qualitative and quantitative approaches. Some of the chapters draw on the close analysis of single genres – e.g. those that explore intertextual chains and hypertextual linkages. Others use corpus-assisted discourse analytical tools to study small-scale corpora of genres to investigate rhetorical, linguistic, discoursal, pragmatic, register and style features of genres. Other chapters illustrate methodological approaches such as multimodal analysis of genres combining visual and verbal modes, or ethnographic analysis to enquire into the producers of those genres, offering 'thick descriptions' (Lillis 2008) of the rhetorical situation in order to provide rich insights into how certain affordances of digital communication account for variation in the linguistic-pragmatic-rhetorical interfaces of Internet genres. Succinctly put, the contributors use different methodological approaches to research genres for the communication of science on the Internet and discuss issues central to the fields of scientific and research communication, genre studies and online communication: the webification of science, multimodality in the communication of scientific knowledge, public communication of science, genre remediation and genre hybridity.

The first two chapters explore the evolution of the research article when imported to the Internet and the changes in the discursive environment that online research articles inhabit. Both articles provide evidence that partly question some researchers' assertion that the migration to the Internet has not led to significant changes in the research article. Harmon uses the Public Library of Science (PLS) to show that, even if the typical Internet scientific article is not very different in form and function from the 20th-century printed version, a revolution in scientific communication may be under way, led by the PLOS and other innovations like Elsevier's "Article of the Future." Harmon first offers a review of studies which have explored how online scientific articles and journals are different from their printed versions. He then presents an analysis of two "enhanced" PLOS articles to illustrate how Internet technologies can bring about changes in how scientific knowledge is represented, communicated and accessed. In doing so, Harmon addresses several of the issues discussed in this volume: the evolution and enhancement of academic genres harnessing digital affordances, the hybrid and multi-generic nature of many academic genres online, the evolution of genres and the emergence of new genres to accommodate wider audiences. Finally, and highly interestingly, he offers his vision on the additional changes that might be expected for digital scientific journals in the coming decades, thus offering pathways for future research on scientific digital genres.

Mehlenbacher and Mehlenbacher posit that the relative stability in form and function of the research article even after migrating to the Internet is due to the fact that it is part of a broader ecology of genres that work together to respond to multiple rhetorical demands. They then present the genre of Registered Report to highlight an important moment of genre innovation and illustrate how the research article is actually evolving to meet new exigences. In a moment where demonstrating the soundness of the method through replication has become increasingly important, the Registered Report changes the peer review process to respond to this exigence. The authors' description of a novel process of peer review, involving two stages (review of the methods presented in the Registered Report and of the results presented in the research article) shows the complex interaction between traditional and new genres.

The chapter by Hendges and Florek examines a specific add-on genre that enhances the online research article by harnessing multimodal affordances: the graphical abstract (GA). Hendges and Florek combine textual analysis and data from authors' answers to an online questionnaire to study graphical abstracts in two disciplines – Chemistry and Engineering. By comparing three features of GAs (layout patterns, nature of the images and originality of the images) in these two disciplines, they provide insights on the evolution of the genre, which was adopted earlier in Chemistry. The analysis of the originality of the images (when compared to the images in the article) in both disciplines reflects that, as the genre evolves, the relation with the genre on which it draws (i.e. the research article) changes and becomes more complex.

In their chapter Rowley-Jolivet and Carter-Thomas show how the discourse of scientific printed genres is remediated in four audiovisual genres available in the digital medium to communicate scientific research to a predominantly non-specialized audience (Three-Minute Thesis presentations, author videos, and podcasts on a popular science and a research journal website). The authors analyze the content, rhetorical structure and communicative strategies in these genres and compare them with the established written genres they are related to (research articles, PhD dissertations). In doing so they address some of the issues which are particularly relevant for the analysis of the relation between digital genres and printed genres of science communication: the erosion of boundaries between expert and public audiences, the capability to reach wider audiences, and the shift to a more conversational and personalized style of science communication triggered by the push for science democratization. Their chapter also raises questions important for genre theory, such as the need to reconsider whether audience can still be deemed as a defining criterion of genre analysis.

The chapter by Breeze explores remediation in relation to peer reviews. Breeze analyzes how changes in the peer review process, moving to an open evaluation system, may have an effect on the ways referees and authors interact and on the strategies used by authors to construct their responses when these are going to be published online. Drawing on the tradition of ESP genre studies and combining content analysis of genre moves and corpus analysis, the author compares open responses to reviewers in *eLife* and responses to reviewers in non-open-peer-review journals. Breeze uses the theory of politeness and relational work to show how the possibility for authors to justify their work in front of the community against public criticism can be an influential factor affecting the evolution from traditional author responses to open author responses. The openness facilitated by Internet affordances seems to lead to changes in the relational work carried out by the authors and the complexity of the genre.

The following chapters (by Maier and Engberg; Mirović, Bogdanović and Bulatović; Orpin; Smart and Falconer; and Reid and Anson) are concerned with the erosion of boundaries between audiences and the relations between genres intended for audiences with different degrees of expertise or disciplinary knowledge. Maier and Engberg adopt a multimodal perspective to explore knowledge mediation and remediation across several media. They all focus on the strategies used when the scientific knowledge constructed in research-based articles is remediated in other generic forms to reach various audiences. In order to do so they propose two conceptual frameworks, knowledge mediation processes and levels of explanatory depth, which aim at describing how various semiotic modes interact in the process of creating textual meaning. Furthermore, they demonstrate the analytical potential of these frameworks by using them to analyze the relations between three articles published in the *Harvard Business Review* and the generic forms accompanying them (an interview, a video and a webinar).

Mirović, Bogdanović and Bulatović also address issues of multimodality, remediation, hybridity and public communication of science. They analyze how highly specialized discourse is recontextualized in different multimedia formats in the online edition of the *IEEE Spectrum* magazine (i.e. videos, podcasts and slide-shows) to meet the information needs of the magazine's semi-expert audience, and how the affordances of different modes contribute to accomplishing knowledge dissemination. The authors draw on the framework of multilevel knowledge building processes (Engberg and Maier 2015; see also Maier and Engberg this volume) to analyze how different media interact to extend, elaborate or enhance knowledge and adapt it to the level of expertise of the audience. The chapter also explores genre hybridity, by examining the integration of multiple genres and different semiotic resources in the magazine website.

In her chapter, Orpin uses corpus analysis complemented with multimodal analysis to examine the rhetorical strategies used to recontextualize scientific information in ECDC (European Centre for Disease Control and Prevention) reports into Twitter discourse. The author discusses how the writers exploit the affordances of the microblog technology to reach and engage a diversified audience (made up of health professionals, public policy officials and the lay public). She also shows how not only the affordances of the medium but also the constraints of the genre (i.e. the limitation in the number of characters, in the case of microblogs) determine how scientific knowledge is recontextualized.

The chapter by Smart and Falconer focuses on the discursive representations of science and technology embedded in the encyclical *Laudato Si'* and examines recontextualization by exploring two genre sets where the encyclical occupies a place, as well as the digital responses triggered by *Laudato Si'*. Smart and Falconer show how the first genre set (the ACTA, the summary report, the papal address and the papal encyclical) has served as a medium for the ongoing exchange of knowledge and beliefs between Academy scientists and eight popes. They also explain how a second digital genre set serves to recontextualize aspects of *Laudato Si'*'s representation of science/technology into digital genres which can reach a broad and diverse audience.

Lastly, the chapter by Reid and Anson gives insight into the public communication of science and into how scientists use digital genres to accommodate public audiences. Starting from the assumption that many digital genres are "polycontextual," i.e. they simultaneously achieve several purposes, reach diverse audiences and are part of multiple networks, these authors explore polycontextuality in the digital writing of a team developing a citizen science project. Their analysis provides a clear illustration of how scientists can both harness purposefully and compose for context collapse: the scientists in the team composed their texts, and particularly the figures, to act in multiple contexts, making them both specific enough for specialized audiences and accessible enough for public audiences. The authors claim that by including public digital genres in their repertoires of scientific communication, researchers can create a different type of scientific community, where citizen amateur-experts can also participate.

It is hoped that the themes foregrounded in this volume shed further light into the recontextualization of expert knowledge through Internet genres and lend a timely relevance to the increasing digital turn of scientists' social activity.

Acknowledgement

This volume is a contribution to project "Ecologies of genres and ecologies of languages: an analysis of the dynamics of local, cross-border and international scientific communication" (FFI2015-68638-R MINECO/FEDER, EU) funded by the Spanish Ministry of Economy and Competitiveness and the European Social Fund. The volume is also a contribution to project "Comunicación internacional y retos sociales" (H16_17R) financed by the Government of Aragon and co-financed by FEDER 2014–2020 "Construyendo Europa desde Aragón."

References

Aalbersberg, IJsbrand Jan, Frans Heeman, Hylke Koers, and Elena Zudilova-Seinstra. 2012. "Elsevier's Article of the Future Enhancing the User Experience and Integrating Data through Applications." *Insights: the UKSG Journal* 25 (1): 33–43. https://doi.org/10.1629/2048-7754.25.1.33

Aalbersberg, IJsbrand Jan, Judson Dunham, and Hylke Koers. 2013. "Connecting Scientific Articles with Research Data: New Directions in Online Scholarly Publishing." *Data Science Journal* 12: 235–242.

Askehave, Inger, and Anne E. Nielsen. 2005. "Digital Genres: A Challenge to Traditional Genre Theory." *Information Technology & People* 18 (2): 120–141. https://doi.org/10.1108/09593840510601504

Ball, David. 2016. The Impact of Open Science. Last accessed June 15, 2019. https://www.fosteropenscience.eu/content/impact-open-science

Bardi, Alessia, and Paolo Manghi. 2014. "Enhanced Publications: Data Models and Information Systems." *LIBER Quarterly* 23 (4): 240–273. https://doi.org/10.18352/lq.8445

Bateman, John A. 2008. *Multimodality and Genre*. Basingstoke, UK: Palgrave Macmillan. https://doi.org/10.1057/9780230582323

Bawarshi, Anis S., and Mary Jo Reiff. 2010. *Genre: An Introduction to History, Theory, Research and Pedagogy*. Anderson, SC: Parlor Press.

Bazerman, Charles. 1994. "Systems of Genres and the Enactment of Social Intentions." In *Genre and the New Rhetoric*, ed. by Aviva Freedman, and Peter Medway, 79–101. Bristol, Pa: Taylor and Francis.

Bazerman, Charles. 1997. "The Life of Genre, the Life in the Classroom." In *Genre and Writing*, ed. by Wendy Bishop, and Hans Ostrum, 19–26. Portsmouth: NH Boynton/Cook.

Bernad-Mechó, Edgar. 2015. "A Multimodal Discourse Analysis of Linking Metadiscursive Elements in Two OpenCourseWare Lectures (MOOCs)." *Procedia – Social and Behavioral Sciences* 212: 61–66. https://doi.org/10.1016/j.sbspro.2015.11.299

Berkenkotter, Carol, and Thomas N. Huckin. 1995. *Genre Knowledge in Disciplinary Communication Cognition/Culture/Power*. Hillsdale, NJ: Lawrence Erlbaum.

Bhatia, Vijay K. 1993. *Analysing Genre — Language Use in Professional Settings*. London: Longman.

Bhatia, Vijay K. 2004. *Worlds of Written Discourse: A Genre-Based View*. London: Continuum.

Biber, Douglas, and Susan Conrad. 2009. *Register, Genre, and Style*, 2nd edition. Cambridge: Cambridge University Press. https://doi.org/10.1017/CBO9780511814358

Biber, Douglas, and Bethany Gray. 2012. "The Competing Demands of Popularization vs. Economy: Written Language in the Age of Mass Literacy." In *The Oxford Handbook of the History of English*, ed. by Terttu Nevalainen, and Elisabeth C. Traugott, 314–328. Oxford: Oxford University Press.

Bolter, Jay D., and Richard Grusin. 1999. *Remediation – Understanding New Media*. Cambridge: The MIT Press.

Bondi, Marina. 2018. "Blogs as Interwoven Polylogues: The Dialogic Action Game." *Language and Dialogue* 8: 43–65. https://doi.org/10.1075/ld.00004.bon

Boulton, Geoffrey. 2012. *Science as an Open Enterprise*, UK: The Royal Society. Last accessed June 15, 2019. http://royalsociety.org/policy/projects/science-public-enterprise/report/

Breeze, Ruth. 2016. "Tracing the Development of an Emergent Part-Genre: The Author Summary." *English for Specific Purposes* 42: 50–65. https://doi.org/10.1016/j.esp.2015.11.003

Breure, Leen, Hans Voorbij, Maarten Hoogerwerf. 2011. "Rich Internet Publications: 'Show What You Tell'." *Journal of Digital Information* 12 (1). Last accessed June 15, 2019. https://journals.tdl.org/jodi/index.php/jodi/article/view/1606/1738

Buehl, Jonathan. 2016. "Revolution or Evolution? Casing the Impact of Digital Media on the Rhetoric of Science." In *Science and the Internet: Communicating Knowledge in a Digital Age*, ed. by Alan Gross, and Jonathan Buehl, 1–10. Amityville, NY: Baywood Publishing.

Carter-Thomas, Shirley, and Elizabeth Rowley-Jolivet. 2016. "Open Science Notebooks: New Insights, New affordances." *Journal of Pragmatics* 116: 64–76. https://doi.org/10.1016/j.pragma.2016.12.003

Casper, Christian. 2016. "The Online Research Article and the Ecological Basis of New Digital Genres." In *Science and the Internet: Communicating Knowledge in a Digital Age*, ed. by Alan Gross, and Jonathan Buehl, 77–98. London: Routledge.

Crawford Camiciottoli, Belinda. 2018. "OpenCourseWare Lectures: A New Genre for Knowledge Dissemination?." Paper presented at the CERLIS Conference 2018, Bergamo, Italy, June 21–23, 2018.

Crowston, Kevin, and Marie Williams. 2000. "Reproduced and Emergent Genres of Communication on the World Wide Web." *The Information Society* 16 (3): 201–215. https://doi.org/10.1080/01972240050133652

Crowston, Kevin and Barbara Kwasnik. 2004. "A Framework for Creating a Facetted Classification for Genres: Addressing Issues of Multidimensionality." *Proceedings of the 37th Hawaii International Conference on System Sciences*, 1–9. Los Alamitos: IEEE Computer Society Press.

Delfanti, Alessandro. 2015. "Preprints and Peer Review: A Lesson from Physics." *Journal of Science Communication*. 14. https://doi.org/10.22323/2.14040305

Devitt, Amy J. 2004. *Writing Genres*. Carbondale, IL: Southern Illinois University Press.

Elsevier. 2018. *About the Audioslides*. Last accessed June 15, 2019. https://www.elsevier.com/books-and-journals/content-innovation/audioslides

Engberg, Jan, and Carmen D. Maier. 2015. "Exploring the Hypermodal Communication of Academic Knowledge beyond Generic Structure." In *Discourse In and Through the Media: Recontextualizing and Reconceptualizing Expert Discourse*, ed. by Marina Bondi, Silvia Cacchiani, and Davide Mazzi, 46–62. Newcastle upon Tyne, UK: Cambridge Scholars Publishing.

Erviti, Mari Carmen, and Erik Stengler. 2016. "Online Science Videos: An Exploratory Study with Major Professional Content Providers in the United Kingdom." *Journal of Science Communication* 15 (6). A06. http://eprints.uwe.ac.uk/30236

Gross, Alan, and Jonathan Buehl (eds.). 2016. *Science and the Internet: Communicating Knowledge in a Digital Age*. Amityville, NY: Baywood Publishing Company. https://doi.org/10.4324/9781315231099

Heyd, Theresa. 2016. "Digital Genres and Processes of Remediation." In *The Routledge Handbook of Language and Digital Communication*, ed. by Alexandra Georgakopoulou, and Tereza Spilioti, 87–102. London, UK: Routledge.

Hyland, Ken. 2011. "The Presentation of Self in Scholarly Life: Identity and Marginalization in Academic Homepages." *English for Specific Purposes* 30 (4): 286–297. https://doi.org/10.1016/j.esp.2011.04.004

Kaplan, Sarah, and Joanna Radin. 2011. "Bounding an Emerging Technology: Para-scientific Media and the Drexler-Smalley Debate about Nanotechnology." *Social Studies of Science* 41 (4): 457–485. https://doi.org/10.1177/0306312711402722

Kelly, Ashley R., and Carolyn R. Miller. 2016. "Intersections: Scientific and Parascientific Communication on the Internet." In *Science and the Internet: Communicating Knowledge in a Digital Age*, ed. by Alan Gross, and Jonathan Buehl, 221–245. Amityville, NY: Baywood.

Kjellberg, Sara. 2009. "Scholarly Blogging Practice as Situated Genre: An Analytical Framework Based on Genre Theory." *Information Research* 14 (3). http://www.informationr.net/ir/14-3/paper410.html

Kuehne, Lauren, and Julian Olden. 2015. "Lay Summaries Needed to Enhance Science Communication." *Proceedings of the National Academy of Sciences of the United States* 112 (12): 3585–3586. https://doi.org/10.1073/pnas.1500882112

Lorés, Rosa. 2018. "Digital Spaces of Knowledge Dissemination and Visibility: Promotional Discourse in European Research Projects Websites." Paper presented at the CERLIS Conference 2018, Bergamo, Italy, June 21–23, 2018.

Luzón, María-José. 2013. "Public Communication of Science in Blogs: Recontextualizing Scientific Discourse for a Diversified Audience." *Written Communication* 30 (4): 428–457. https://doi.org/10.1177/0741088313493610

Luzón, María-José. 2017. "Connecting Genres and Languages in Online Scholarly Communication: An Analysis of Research Group Blogs." *Written Communication* 34 (4): 1–31. https://doi.org/10.1177/0741088317726298

Luzón, María-José. 2018. "Engaging with Multiple Audiences in Research Group Websites: The Interplay between Genres and Languages." Paper presented at the 36th International Conference of the Spanish Society for Applied Linguistics, Cádiz, Spain, April 19–21, 2018.

Luzón, María-José. 2019. "Bridging the Gap between Experts and Publics: The Role of Multimodality in Disseminating Research in Online Videos." *Ibérica, Journal of the European Association of Languages for Specific Purposes* 37(1): 167–192.

Mahrt, Merja, and Cornelius Puschmann. 2014. "Science Blogging: An Exploratory Study of Motives, Styles, and Audience Reactions." *Journal of Science Communication* 13 (3): A05.

Marwick, Alice, and Danah Boyd. 2011. "I Tweet Honestly, I Tweet Passionately: Twitter Users, Context Collapse, and the Imagined Audience." *New Media & Society* 13: 114–133. https://doi.org/10.1177/1461444810365313

Mehlenbacher, Ashley R. 2017. "Crowdfunding Science: Exigencies and Strategies in an Emerging Genre of Science Communication." *Technical Communication Quarterly* 26 (2): 127–144. https://doi.org/10.1080/10572252.2017.1287361

Miller, Carolyn R. 1984. "Genre as Social Action." *Quarterly Journal of Speech* 70: 151–167.

Miller, Carolyn R., and Dawn Shepherd. 2009. "Questions for Genre Theory from the Blogosphere." In *Genres in the Internet: Issues in the Theory of Genre*, ed. by Janet Giltrow, and Dieter Stein, 263–290. Amsterdam: John Benjamins. https://doi.org/10.1075/pbns.188.11mil

Miller, Carolyn R., and Jeanne Fahnestock. 2013. "Genres in Scientific and Technical Rhetoric." *Poroi* 9 (1): 12. https://doi.org/10.13008/2151-2957.1161

Nielsen, Michael. 2011. An Informal Definition of Open Science. The OpenScience Project. Last accessed June 15, 2019. http://www.openscience.org/blog/?p=454

Orlikowski, Wanda, and JoAnne Yates. 1994. "Genre Repertoire: The Structuring of Communicative Practices in Organizations." *Administrative Science Quarterly* 39 (4): 541–574. https://doi.org/10.2307/2393771

Paltridge, Brian. 1995. "Working with Genre: A Pragmatic Perspective." *Journal of Pragmatics* 24 (4), 393–406. https://doi.org/10.1016/0378-2166(94)00058-M

Pérez-Llantada, Carmen. 2013. "The Article of the Future: Strategies for Genre Stability and Change." *English for Specific Purposes* 32 (4): 221–235. https://doi.org/10.1016/j.esp.2013.06.004

Pérez-Llantada, Carmen. 2016. "How is the digital medium shaping research genres? Some cross-disciplinary trends." *ESP Today* 4: 22–42.

Plastina, Anna F. 2017. "Professional Discourse in Video Abstracts: Re-articulating the Meaning of Written Research Article Abstracts." In *Specialised and Professional Discourse across Media and Genres*, ed. by Giuliana Garzone, Paola Catenaccio, Kim Grego, and Roxanne Doerr, 57–74. Milano: Ledizioni.

Prior, Paul. 2009. "From Speech Genres to Mediated Multimodal Genre Systems: Bakhtin, Voloshinov, and the Question of Writing." In *Genre in a Changing World*, ed. by Charles Bazerman, Adair Bonini, and Debora Figueiredo, 17–34. Fort Collins, CO: WAC Clearinghouse; West Lafayette, IN: Parlor Press.

Riboni, Giorgia. 2018. "Vlogging Science: Scholarly Vlogs as a New Knowledge Dissemination Genre." Paper presented at the CERLIS Conference 2018, Bergamo, Italy, June 21–23, 2018.

Ross-Hellauer, Tony. 2017. "What is Open Peer Review? A Systematic Review." F1000 Research 6: 588. https://doi.org/10.12688/f1000research.11369.2

Ruiz-Madrid, Noelia, and Inmaculada Fortanet-Gómez. 2017. "An Analysis of Multimodal Interaction in a Webinar: Defining the Genre." *EPIC Series in Language and Linguistics* 2: 274–282. https://doi.org/10.29007/p8mm

Sidler, Michelle. 2016. "The Chemistry Liveblogging Event: The Web Refigures Peer Review." In *Science and the Internet: Communicating Knowledge in a Digital Age*, ed. by Alan G. Gross, and Jonathan Buehl, 99–116. London: Routledge.

Smart, Graham. 2016. "Discourse Coalitions, Science Blogs, and the Public Debate on Global Climate Change." In *Genre and the Performance of Publics*, ed. by Anis S. Bawarshi, and Mary Jo Reiff, 157–177. Logan, Utah: Utah State University Press. https://doi.org/10.7330/9781607324430.c008

Smith, Alison. 2015. "'Wow, I Didn't Know that Before; Thank You': How Scientists Use Twitter for Public Engagement." *Journal of Promotional Communications* 3 (3): 320–339.
Spicer, Scott. 2014. "Exploring Video Abstracts in Science Journals: An Overview and Case Study." *Journal of Librarianship and Scholarly Communication* 2 (2): eP1110. https://doi.org/10.7710/2162-3309.1110
Spinuzzi, Clay. 2004. "Describing Assemblages: Genre Sets, Systems, Repertoires, and Ecologies." *Computer Writing and Research Lab. White Paper Series. [Online].* Last accessed June 15, 2019. http://www.dwrl.utexas.edu/sites/www.dwrl.utexas.edu/files/assemblages.pdf
Swales, John M. 1990. *Genre Analysis: English in Academic and Research Settings.* Cambridge: Cambridge University Press.
Swales, John M. 2004. *Research Genres. Explorations and Applications.* Cambridge: Cambridge University Press. https://doi.org/10.1017/CBO9781139524827
Trench, Brian. 2008. "Internet: Turning Science Communication Inside-out?" In *Handbook of Public Communication of Science and Technology*, ed. by Massimiano Bucchi, and Bruce Lewenstein, 143–186. London, U.K.: Routledge.
Wickman, Chad. 2016. "Learning to "Share your Science": The Open Notebook as Textual Object and Dynamic Rhetorical Space." In *Science and the Internet: Communicating Knowledge in a Digital Age*, ed. by Alan G. Gross, and Jonathan Buehl, 11–22. London: Routledge.
Yeo, Sara K., Xuan Liang, Dominique Brossard, Kathleen M. Rose, Kaine Korzekwa, Dietram A. Scheufele, and Michael A. Xenos. 2017. "The Case of #arseniclife: Blogs and Twitter in Informal Peer Review." *Public Understanding of Science* 26 (8): 937–952. https://doi.org/10.1177/0963662516649806
Yang, Wenhsien. 2017. "Audioslide Presentations as an Appendant Genre – Key Words, Personal Pronouns, Stance and Engagement." *ESP Today* 5 (1): 24–45. https://doi.org/10.18485/esptoday.2017.5.1.2

CHAPTER 2

At the frontlines of the online scientific article

Joseph E. Harmon
Communications and Public Affairs, Argonne National Laboratory

Inaugurated in 1665, the scientific journal article slowly evolved an elaborate set of discourse norms related to style, presentation features like references and figures, and argument. With the recent "webification" of the scientific literature, these norms have remained in place while the digital opportunities afforded by webification have supplemented and enhanced some of them. Leading the way on these fronts has been a set of digital-only scientific journals published by the Public Library of Science (PLOS) – founded in 2000 and now one of the major digital innovators in scientific publishing. This chapter focuses on the substantive changes PLOS and a few other publishers have brought to the genre of both the scientific article and journal, and speculates on what additional changes in the genre might be expected in the coming decades.

Keywords: Public Library of Science, webification, genre change, online scientific article, scientific research article

1. Introduction

Scholars apply the category "genre" to modes of human communication marked by distinct similarities, or as Miller (1984: 152) has defined it for written communication, "a recurring rhetorical situation." Similarly, Giltrow and Stein (2009: 1) define genre as communications in which "features of language aggregate in recognizable patterns." Recent digital examples could include the tweet, blog, and Facebook page. By the above broad definition, the scientific journal article would seem to qualify as a genre, both the print and digital versions. Even though the genre was established in the late 17th century, it seems fair to contend that it is more robust now than it ever has been and is deserving of special scrutiny, especially given the obvious importance of science and technology to modern society.

This chapter begins with a review of past book-length quantitative studies on the scientific article from its origin in 1665 to the present Internet age. These studies have provided many insights into the "recurring rhetorical situations" and

https://doi.org/10.1075/pbns.308.02har
© 2019 John Benjamins Publishing Company

"recognizable patterns" found in the typical scientific article as it has evolved over the centuries. This chapter then examines several exceptional online scientific articles and journals that exploit the new features afforded by the Internet in the hope of catching a glimpse of what one day might become genre conventions. Worth noting is that it is not only bundles of traditional linguistic features that are the primary carriers of change here, but also visual features and hyperlinks to additional verbal and visual information possible to include because of the essentially unlimited display space the Internet allows.

2. Origin and evolution of the print scientific article

The question a number of scholars has investigated with respect to the scientific article is, how have the communicative conventions in this genre changed over time? There have been four book-length semi-quantitative studies on that subject spanning the scientific literature from the 17th to 20th century: first out of the gate being Bazerman's *Shaping Written Knowledge: The Genre and Activity of the Experimental Article in Science* (1988), followed by Atkinson's *Scientific Discourse in Sociohistorical Context: The Philosophical Transactions of the Royal Society of London, 1675–1975* (1999), Gross et al.'s *Communicating Science: The Scientific Article from the 17th Century to the Present* (2002), and Banks's *The Development of Scientific Writing: Linguistic Features and Historical Context* (2008).

The first definitive and thorough study of scientific communication, Bazerman's book, deserves official designation as a "landmark." It subjects to quantitative analysis 100 experimental articles from *Philosophical Transactions* over the period 1665 to 1800 and 40 articles from *Physical Review* over the period 1893 to 1980. Bazerman interprets his data in terms of the social driving forces he views as shaping written knowledge over time in the sciences: in particular, the evolving specialized communities of science, the evolving organizational and economic circumstances within those communities, and new communicative technologies related to printing and illustration. From the acquired data he was able to propose general trends in writing style for both the early and modern scientific article and the forces that shaped them. In addition, by close reading of a *Philosophical Transactions* article (1672) on optics by Isaac Newton and the published critical response to it he tracks how the process shaped the final argument in Newton's *Optics*, published in 1704. He similarly tracks how the communal endeavor of modern physics shaped the many drafts by physicist Arthur Holly Compton in his classic *Physical Review* paper (1923) on the scattering of X-rays by electrons, as well as its post-publication reception. Continuing the work of Bazerman on *Philosophical Transactions*, Atkinson and Banks examined a large selection of articles from its founding in 1665 to 1975, the eve of the shift to digital publication. Atkinson emphasizes the rhetorical changes and socio-historical forces of change, and Banks concentrates on linguistic changes, mainly use of passive voice, first-person pronouns, and nominalizations. Many other scholars have investigated the special linguistic register of scientific discourse in much narrower historical periods (e.g. Barber 1962; Swales 1990; Biber and Jones 2005: Giltrow 2011).

Gross, Harmon and Reidy (2002) quantitatively analyzed scientific texts published between 1665 and 1995. In contrast to other similar studies, they examined a much larger sample (870 full articles plus about 4500 randomly selected ten-line passages) in three languages from a wide assortment of significant journals (English, French, and German). One of the more striking conventions of the 20th-century scientific article identified by Gross et al. is an elaborate finding system of headings, graphic legends, numbered citations, and numbered equations. This master finding system allows scientist-readers to navigate more easily among the diverse components of the article; they can thus extract the desired bits of theory, methods, results, relevant citations, and conclusions without necessarily having to read the text from front to back. Gross et al. also tracked the evolution of the modern article's standard arrangement of parts to form a consistent argument. These parts are an abstract that summarizes the article content in a short paragraph, an introduction that establishes a research problem and its broader context within past research, methods devoted to solving the problem, results and discussion from having applied those methods, and essential conclusions from the result and discussion, along with possible future research directions, followed by acknowledgments of financial and other support along with citations to past research. This modular arrangement acts like a template allowing readers to easily direct their attention on select components within the argument being made. It thereby permits scientists to adjust their reading time to either just get the gist or delve deeply into the technical details. The conventions in the 20th-century scientific article shift as well to much greater prominence given to visual expression. Dominating the visuals are tables that display data and explanations in neat columns and rows and graphs that depict data trends and causal mechanisms. Also of importance are schematics that summarize theoretical constructs and photographs and drawings that capture objects of nature and the research lab. Indeed, as established in *Science from Sight to Insight* (2013) by Gross and Harmon, it is the interaction of visual and verbal texts that constitutes the heart of current communicative practices in the modern scientific article (see also Hendges and Florek this volume). As discussed below, the widespread digitization of the scientific literature is impacting all these trends to varying degrees.

The next section reviews several previous in-depth studies of the current conventions of the digital scientific article as genre.

3. Current state of the digital scientific article

The first extensive study of the online scientific article appears in Owen's *The Scientific Article in the Age of Digitization* (2007). Owen reported the results from his analysis of a large sample of web-only scientific journals (86 of them), mostly starting up in the late 20th century and ending in 2004. Somewhat surprisingly, he concludes that contrary to overblown expectations about the effect of the Internet on scientific communication, the digital scientific article "remains a digital copy of the printed form" (11). Owen views the current situation as preserving and extending "existing functions and values rather than as an innovation that radically transforms a communicative practice that has evolved over the centuries" (55). That seems a fair assessment of the state of Internet affairs at the time for run-of-the-mill online scientific journal from the late 20th century.

In one of the chapters in *The Internet Revolution in the Sciences and Humanities* (2016), Gross and I reported a study of a random sample of 100 articles in ten highly cited online scientific journals and 50 articles from five online-only scientific journals from the year 2012. In our analysis, we noted a major shift in the experience of reading journals and the articles. Readers can not only read pages from left to right, top to bottom; they can also point, click, and scroll to access a host of related materials (databases, other articles, author biographies, etc.), rotate 3D structures, interact with graphs and tables, zoom in and out of maps and structures, watch videos of computer simulations and events as they occurred in the laboratory or field, listen to authors talk about their research, share articles they like with friends, and comment upon claims they question. The printed page has been transformed into a digital page linked to a constantly expanding world-wide hypertext library of scientific materials – visual, verbal, and oral. As one small example, by clicking through links, readers can move backward in time from an article to the articles it cites and forward to the articles that cite it. Scientific articles are thus no longer what Weinberger (2012) refers to as "stopping points" in his description of print documents generally – where all but the most highly motivated readers are discouraged from fact checking or pursuing additional reading or research because of the extra time and effort involved.

The linking possible with the web page in the 21st century has resulted, at least in some journals, in modifications on the standard structure. In these journals, the main text conveys minimal methods information and only essential figures and tables to forward the authors' argument, while a link to a supplemental information section presents additional information in the form of tables and figures, videos and animated visuals, detailed methods, all the data acquired during the research, and further discussion of results. This supplemental information is aimed at the relatively few highly specialized readers who seek more technical details than those provided in the main text. In addition, a side panel offers links to other articles that might be of interest to the readers based on the sort of computer algorithm employed by Amazon.com.

We also reported a small but not insignificant difference between the print and Internet article: metrics relevant to community response to a given article are often available. Because the number of viewers, citation counts, blog posts, and social bookmarking are immediately accessible for a given article, readers have access to quantitative information with which to roughly assess the content of an article and the community judgment of its value. Reader reaction to articles is also possible. Most online journals allow for and immediately publish reader comment and author response, either as part of social media or directly linked to the published text. However, in marked contrast to the reviews of products in Amazon.com, we found such commentary to be anemic in the online literature.

Finally, we noted changes in journals as a whole. Journal contents have been radically transformed into a diverse assemblage of genres. Online journals are publishing not only the usual research articles, editorial essays, and letters to the editor about published articles, as in the past, but also educational materials for students, scientific blogs about original research, peer reviewer comments, databases like the Protein Data Bank, author interviews, and videos about research articles of interest to the public.

Our findings suggested that the scientific journal article as a genre, at least as evidenced by the leading journals, is undergoing rapid and substantive changes. It should be emphasized that, consistent with Owen's findings, much about the digital scientific article remains unchanged from the genre of the 20th-century printed version – the Internet era has not overseen a complete makeover. By and large, the standard structure remains firmly intact and even more homogenized, with the exception of the inclusion of supplemental information at the end that is now fairly routine. The writing style remains heavily passive voice and cognitively dense, with technical information packed into noun phrases teeming with multiple modifiers (before and after the central noun), quantifications, and abbreviations. First-person pronouns appear about once every hundred words, and purposely literary language is largely absent. The density of figures and tables within the main text also has not noticeably changed, and the inclusion of videos or interactive images is still an exception, not a norm. On the basis of our analysis of the elite online scientific literature, we concluded that

> The web does not redefine the essence of the scientific article; it has always been a network of knowledge. Rather, the web permits the exploration and exploitation of possibilities inherent in the structure of scientific communication. The article is what it always was, only more intensely, more completely. It is too soon to tell which variations now evident on the web will prove enduring. But while scientific

> argument remains as rigorous as ever, the means of representing it, providing evidence for it, and accessing it will never be the same.
>
> (Gross and Harmon 2016: 51)

Another assessment of the current state of online scientific communication appears in the collection of essays edited by Gross and Buehl, *Science and the Internet: Communicating Knowledge in a Digital Age* (2016). Many of the essays have a bearing, direct and indirect, on the scientific journal article as a genre. For example, Wickman (2016) discusses the transformation of the genre of the laboratory notebook. By means of the Internet, scientists can share their research in the making for the purpose of critical evaluation or even adaption for new research projects by others in the scientific community. In his essay, Gross (2016) contends that the Internet is transforming the genre of investigative science journalism in the form of web sites that publish claims of falsification of data, plagiarism, and even major inadvertent errors after scientific article publication. In separate essays, Buehl (2016), Casper (2016), and Sidler (2016) show how the Internet is intensifying and speeding up critical evaluation of the scientific literature in blogs and commentary space built into current scientific journals. And Fahnestock (2016) offers an important corrective to all the good news being reported about science communication and the Internet: the distortion of the scientific literature that appears to be routinely happening when re-interpreted and misinterpreted by non-experts in blogs and online news type articles: "With its interactivity, the Internet has amplified uninformed responses to various research reports, impeding the creation of an informed public" (Fahnestock 2016: 117).

Here we arrive at the central question of this chapter: What is the future of the digital scientific article as a genre? Answering that question does not lend itself to the quantitative analyses in such studies as Owen (2007), Harmon (2016), and Gross and Harmon (2016). Such analyses can only provide reliable insight into present practices. In the next section (4.1), I begin with a deep dive into the advanced communicative practices of a single article published by the Public Library of Science (PLOS), selected for close analysis because it embodies many of the occasional communicative practices that take advantage of Internet publication – practices that could become norms one day. The next section (4.2) examines two other PLOS articles, each of which exemplifies some advanced communicative features afforded by the Internet.

It is important to bear in mind that PLOS is by no means the only scientific publisher innovating with the Internet. The penultimate section brings to the fore advanced communicative practices found in other journals. The overall aim is for the reader to come away with a concrete sense of the communicative practices that could become norms in the future scientific journal based on examples at the current cutting edge.

4. Future of the digital scientific article

At the time of the launch of its first journals in 2003, PLOS had as its mission to "lead a revolution in scientific communication." From the start, that has meant the creation of peer-reviewed online-only journals that provide free and unrestricted access to their articles, along with no fee for reproduction of figures and tables. To date, PLOS has published approximately 165,000 open-access articles from 190 countries in eight journals.

All these journals have respectable citation metrics, the latest counts being available through the PLOS web site. As previously mentioned, Gross and I spotted some general trends as determined from a semi-quantitative analysis of a random sample of articles from PLOS as well as the most highly cited scientific journals in the world (Gross and Harmon 2016: 17–51; Harmon 2016: 33–58). But the full impact of Internet-related changes is best appreciated with specific examples – not examples of a typical PLOS article or typical Internet scientific article in general, but examples in which the article makes exceptional use of the new communicative opportunities the Internet permits. The assumption is that by a close reading of such exemplary articles we might thereby better glimpse the future or at least a possible future. To that end, an analysis of the contents of two PLOS articles follows: a state-of-the-art article from *PLOS Biology* (Hobaiter et al. 2014) and a one-of-a-kind, "enhanced" article from *PLOS One* (Karlberg et al. 2009). For convenience, I will be using the terminology "print genre," meaning scientific articles from the days before the Internet, and "online genre," meaning 21st-century scientific articles displayed on the Internet.

4.1 State-of-the-art PLOS article

For centuries, the print journal has been periodically published as single issues containing multiple scientific articles and related items like book reviews, loosely connected to each other but normally consistent with the journal title. The coming of the online journal has weakened the significance of a discrete journal "issue" published on a set date, but not eliminated it entirely. While all but one of the PLOS journals publish periodic issues electronically, all publish articles almost immediately after acceptance and before official publication in an "issue," speeding up the publication process. Further diminishing the importance of journal issues is the likelihood that most readers access specific online articles by Internet searching, not by consulting an individual issue or volume of issues as in the print past.

The online article selected for scrutiny here is featured on the "cover" to the September 2014 issue of *PLOS Biology*. This cover more or less simulates the print genre. It consists of a headline with the article title, below which appears a catchy picture with caption. This particular cover picture depicts a wild chimpanzee eating some plant-like substance. As clearly conveyed in the article title, "Social Network Analysis Shows Direct Evidence for Social Transmission of Tool Use in Wild Chimpanzees," the authors are first emphasizing the method by which they arrived at their claim for new knowledge, social network analysis. By contrast, the picture and its caption emphasize the following phrases in the title, that is, the cultural transmission of tool use in wild chimpanzees – in this case, the creation of a mossy sponge with leaves to drink water from a small watering hole. The title is meant to appeal to fellow researchers in the field; the picture and caption, to all who might be interested in the subject matter. This dual implied audience is in keeping with *PLOS Biology* being freely available to one and all – not just to journal subscribers or those with access to a well-stocked research library – and also being discoverable in serendipitous or focused Internet searching, completely independent of the journal issue. Indeed, after the cover disappeared from the *PLOS Biology* home page with replacement by the cover for the next issue, readers had to access the September 2014 issue as a whole by browsing via a link on the journal's home page or including "September 2014" in an Internet search with other key words.

4.1.1 *Front matter*

The article title at the head of the web page is really no different from what one can easily find in the print genre. What is different is its surrounding electronic environment. To the right of the title (see Figure 1), for example, are metrics with which readers can gauge community response to the article: saves, citations, views, and shares. The metrics in this case suggest robust interest within the scientific community (over 27,000 views and 66 citations as of July 2018).

Beneath the title is a row of four shaded grey boxes that can be opened by clicking. The first has information about the authors, including more information than typically made available in the print genre: not only their names and affiliations, but also a statement of competing interest to rule out possible bias in gathering and interpreting data and an account of who did what in the research. The latter has become necessary as author lists have steadily grown in the Internet era and participation in a research project become ever more compartmentalized (Wuchty, Jones, and Uzzi 2007; Jones, Wuchty, and Uzzi 2008).

The second box visualizes the article metrics, and the third displays any written comments made online about the article. This article has three comments – none of much substance. In general, the comments section of the online genre has yet to become a site of serious dialogic engagement between authors and their audience, though there are online forums being created in an attempt to stir up debate on controversial topics, such as PLOS's *Responding to Climate Change* (https://channels.plos.org/rtcc).

Figure 1. Opening to state-of-the-art PLOS article (Hobaiter et al. 2014)

The final box, "Related Content," provides insight into the reception of this article outside the scientific community. Within this box, the hyperlinked long list of media mentions (21 items) includes articles from *BBC News*, *NBC News*, *The Scientist Magazine*, *New York Times*, *Nature News and Views*, and *The Telegraph*, as well as blog coverage. In general, the titles accurately and vividly capture the main conclusions in the research article: "Chimps with Tools: Wild Ape Culture Caught on Camera" (*BBC News*), "Chimp Culture Caught on Camera" (*Scientist Magazine*), "Chimps Show a Thirst for Learning" (*New York Times*), and "Chimpanzees Develop Cultures in the Wild Like Humans" (*Telegraph*). While many of these news articles emphasize the videotaping and photographing of the chimpanzees in the process of learning how to use leaf sponges for drinking from a small watering hole, as we shall see, that bit of observational evidence is buried deep in the article's Supplemental Information. Also central to the news articles are quotations from the scientist-authors about their findings, who emphasize the observational aspect of their research in the wild. By contrast, as evident in the article itself, the community of scientists values methods and measurements over observational videos or photographs or personal testimony.

Also displayed in the same "Related Content" box is a kind of interactivity and intertextuality that is unique to the online genre: recommended reading – if you like this article, you might want to read these other scientific articles, available by clicking on a link. Recommended reading under select topics is also available on the far right column on the page, as are the over 200 tweets related to the article from various readers. Together, these upfront links to related scientific articles, media coverage, blogs, reader comments, and tweets illustrate Casper's assertion that the online research article has evolved into a blend of different genres and "the core around which a cluster of texts arises" (2016: 96).

Also a product of the expanded page space afforded in the online genre, situated immediately below the row of linked boxes, is the inclusion of two abstracts for the article – one for specialist readers, the other for non-specialists. The first abstract is typical of what one finds in any scientific article from the early 20th century onward – a short paragraph on what the authors did, how they did it, and what they discovered. The problem as stated in this abstract is that the technique of "social network analysis of wild populations" mentioned in the title, as previously practiced, is unduly limited in scope. From the abstract, we learn that the authors modified and improved social network analysis and then used it for studying the social behavior of a Sonso chimpanzee community in the Budongo Forest, Uganda. Their discovery was that their revised method allowed them to obtain, for the first time, direct experimental evidence of social learning by chimpanzees in the wild, to be specific, "a 15-fold increase in learning rate for each time a novice observed an individual moss-sponging."

Piggybacking on this shorter abstract is what PLOS calls an "author summary" aimed to the level of an undergraduate student or science journalist, a type of abstract new to the specialized scientific literature. Whereas in the specialist abstract the problem was a methodological one that is of concern to the research community, in this non-specialist abstract the problem shifts to the lack of any experimental evidence of social learning in chimpanzees in the wild – considered by the scientific community as the most "cultural" of all animals aside from homo sapiens – presumably a problem of interest to a wider audience than those curious about a methodological problem. When discussing how to solve the problem in this abstract, the authors then shift to their improved method and the specific subjects studied, wild chimpanzees in Uganda.

In both types of abstracts, the authors close with a statement regarding the wider implications of their research – their direct evidence for social learning among chimpanzees suggests that "culture originated in a common ancestor of great apes and humans, long before the advent of modern humans." That is what attracted so much media coverage for the article as well as attention from within the scientific community.

Another area of major change in the online genre concerns the visualization of science (see also Hendges and Florek, Rowley-Jolivet and Carter-Thomas, and Mirović, Bogdanović and Bulatović this volume). This is immediately evident by the inclusion, directly below the two abstracts, of thumbnail versions of the figures and tables in the main text. These visualizations display the data acquired from the author's application of their modified method for social network analysis. So with the abstracts and visuals at the front, specialist readers already have at their fingertips all the key information they need about the contents to follow, and are thus better informed than ever about whether it might be worth their while to read onwards.

In PLOS journals, the spatial freedom permitted by online publication has engendered not only two abstracts but a page's worth of what might be called "metadata," ancillary information related to the main text, a section not found in the print genre. For this particular article, that includes the full citation, key dates (received, accepted, published), name of the "academic editor" assigned to oversee the peer review process, copyright information, link to the complete data for the article, funding sources, competing interests, and definitions of acronyms. The link to the collected data is especially noteworthy. An important and substantive change from the print genre is that readers can routinely access the complete data set used during the research. In the print past, the scientific article presented an argument for some new knowledge claim typically supported with data carefully selected and presented to substantiate that claim. The online link to the complete data set in PLOS articles like that of Hobaiter et al. works as a counterargument to any reader who might claim the authors cherry-picked their data, or incorrectly analyzed it, or even may have fabricated some of it out of thin air (although that is still, of course, a possibility).

4.1.2 *Main text*

In contrast to the front matter, the main article that follows looks very much like a typical print version from the 20th century. However, the prose style does show influences from different centuries. As mentioned earlier, the 21st-century readers of online articles like Hobaiter et al.'s obviously do not turn pages, but scroll and point and click on links that tempt them to immediately consult references, specific figures or tables, and different sections within the main text. This makes for a very different reading experience, where readers are purposely discouraged from reading beginning to end, and encouraged to jump from, say, the abstract at the beginning to a news article or a comment by a reader, or maybe to the methods or conclusion at the end.

The overall structure of the main text is typical of the print genre from the 20th century, the main components being introduction, results, discussion, and methods. But in the introduction to this particular article, there is some evidence of a subtle change in content that may be related to the above-mentioned change in reading method. The typical 20th-century introduction establishes a research area, a niche or problem within that area, and only a slight hint at the solution. This introduction has all three elements but offers far more than a hint at the solution.

Instead, there is a summary of the solution and an argument for its validity. For example, we learn in the introduction that "During 6 d of continuous observation, various individuals were observed to develop two tool behaviors, novel to the group: 'leaf-sponge re-use' and 'moss-sponging'. Both behaviors spread partially through the group (Figure 1 and 2)." The displayed figures (not reproduced here) are maps of the social interactions among the chimps leading to the two behaviors. The titles of each have links to databases with the results from the research. And the final long paragraph to the introduction provides the authors' rebuttal anticipating the main counterargument to their conclusion: that these behaviors had a purely environmental explanation as opposed to social learning. With this added information to the introduction it becomes more of a mini-paper than a typical 20th-century introduction, making it another possible stopping point for many readers.

In this paper, the results section following the introduction displays a shift in tone, at times hearkening back to what one can find in a 17th- or 18th-century observational articles in the natural sciences. We are given a brief verbal picture of what the scientists observed: "Our analysis starts with the alpha male NK extracting water from the waterhole and fabricating a moss-sponge (M, November 14, 2011; 9:05 a.m.), while being observed by the adult dominant female NB. Over the following 6 d period, the waterhole was revisited regularly and we observed a further seven individuals fabricating and using moss-sponges (M)." But we also find an interweaving of the key statistical data typical of the quantitative style more typical of the 20th and 21st centuries: "Though we cannot rule out social transmission of RU1 (see Table 2), effects weaker than for M ($\Sigma w_i = 0.246$; compared with the same social effect, $\Sigma w_i = 0.0003$; Table 1)." Consistent with the 20th-century article, the discussion following the results presents the main claims being made from the results, counterarguments to possible alternative explanations or criticisms, and future directions for continued research.

A detailed and lengthy methods section follows the discussion. Although logically the methods belong before the results from having applied the methods, starting in the 20th century, many journals have placed them at the end of the main text because those details are typically only of interest to a few specialist readers. In the online genre, some journals are now relegating the bulk of this section to the Supporting Information (discussed next), which is typically only accessible through the Internet. But such is not the case in this PLOS journal, where the methods section at the end is fairly typical of the 20th century. However, one feature that is slightly different is that, as part of this particular methods section, the authors define key terms in the main text: "leaf sponge," "moss sponge," and "fabrication," "use," and "re-use" of one of those two items. And to better illustrate the terms, they point their readers to the four video recordings in the Supporting

Information of chimpanzees caught by camera in the wild, the inclusion of videos being a key addition to the online genre.

4.1.3 *Supporting information*

This section is entirely new to the scientific journal in general, is usually not included in the printed issues, and typically includes methods details, additional figures and tables of data, and any videos or interactive graphics. The Supporting Information in this specific PLOS example includes seven figures, four tables of data, and four movies, but no methods details other than those incorporated into some of the figures and tables. In keeping with the PLOS policy of open access, all these figures had been uploaded to Figshare, a large communal online database for sharing and using any research output, from figures and posters and PowerPoint presentations to datasets and computer code. The first two figures are photos of the experimental site, absent the subjects, and are supplemental to the verbal observations of it in the results discussion. The next four figures are further visualizations of the acquired data, as are the four tables – all typical of 20th-century visual practice.

While barely mentioned in the main text, the four videos of chimps learning to use a new tool for drinking from a waterhole are what attracted much of the media attention, as evident from the headline "Wild Ape Culture Caught on Camera" (*BBC News*). In 1985, Steven Shapin and Simon Schaffer introduced the concept of "virtual witnessing" in scientific writing related to the 17th-century scientist Robert Boyle. In part, this involved Boyle's use of extensive circumstantial details in written observations, whether experimental or in the field. The online video has brought virtual witnessing to a whole new level. The first video shows chimp NK gathering moss from the tree truck and adding some leaves to it while NB watches. In the next video, NB gathers some moss and adds it to her leaf sponge while being observed by chimp HL, who repeats that action. The next video shows young chimp KZ picking up a leaf sponge from the ground, then using it at the waterhole. The last video shows another young chimp, KS, extracting a leaf sponge from his mother's mouth, then using it at the waterhole. With the coming of online videos like these, readers can for the first time see what the scientists saw while conducting their research. It seems reasonable to predict that such enhanced visuals will have a larger and more prominent space in the future online article.

4.2 Enhanced PLOS articles

In the period 2009 to 2012, PLOS began testing what it called "enhanced" digital articles in several biochemical articles, articles that go somewhat beyond the state of the art as represented by the 2014 Hobaiter et al. article discussed above.

Driving one of the innovations in visualization is three-dimensional, manipulable representations of structural data on a large biomolecule by computer. While such images are not typical, they are not unusual in the present online scientific literature (for one example, see Klinkhamer et al. 2017). But a one-of-a-kind 2012 article by Karlberg and colleagues experimented with a new twist on the display of such images (Figure 2).

The enhancement mainly concerns the splitting of the screen into two distinct components. On the left side is the typical PLOS digital article with embedded figures as thumbnail images. On the right side is a 3D interactive image of the central biomolecule under discussion, which is identified in the article's title, "Crystal Structure of the ATPase Domain of the Human AAA+ Protein *Paraplegin/SPG7*" (italics added). It shows the ribbon crystal structure of the human protein under investigation. Should we so desire with the online version, we can investigate the visual on our own by rotating it in any direction, zooming in and out, and translating it into different kinds of structural representations. Given the infinite number of views of the biomolecule we can create, if we were experts in the field, we might even find something important there not seen by the authors.

A short historical survey of the use of visuals in the scientific article can help put this Karlberg et al. enhancement into perspective. In the 17th and 18th centuries, a leading convention of the print genre was that it provided verbal arguments for claims to new knowledge. The inclusion of visuals was an option and often relegated to the end of the journal issue in a section with all the other figures from that issue. In the following centuries, the visual became integrated into the text itself and steadily grew in importance to the point where it is now a convention of the genre. In the online scientific article, the verbal and visual components hold equal status, and the readers can see for themselves (and hear sometimes as well) in ways never before possible.

In effect, the enhancement to the Karlberg et al. article creates something akin to a left brain (primarily verbal)/right brain (primarily visual) model. The "left brain" is linked to the "right brain" by periodic snatches of text on the left side of the screen that appear throughout the article. When the reader clicks on these links, the visual on the right rotates or zooms in to illustrate the point being made in the text. For example, the key sentence in the abstract is linked, "We present the crystal structure of the AAA-domain of human paraplegin bound to ADP at 2.2 Å." Clicking on the underlined link rotates the molecule to reveal the crystal structure in the form of the labeled interweaving ribbons shown in Figure 2. A later link in the text on the left zooms in to reveal the "ADP" (adenosine diphosphate, an essential ingredient in the flow of energy in biological cells) within that crystal structure on the right. With this manipulable image, this enhanced article is even more interactive than the highly interactive typical PLOS article.

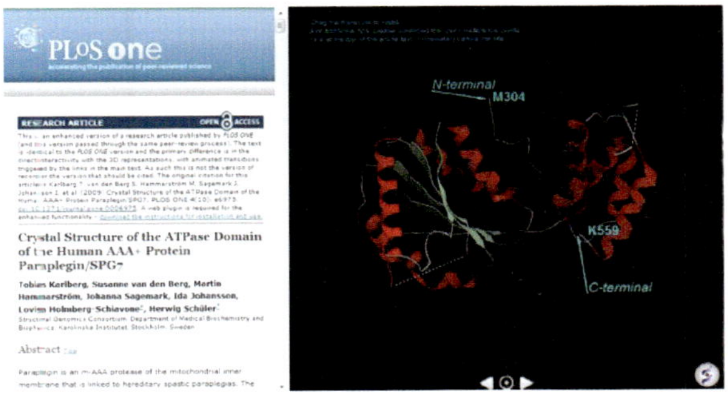

Figure 2. Opening to enhanced PLOS article (Karlberg et al. 2012). See video at http://global.oup.com/us/companion.websites/9780190465933/videos/ch2/video2.4

A different possible genre innovation explored by PLOS is semantic markup. As noted by Shotton et al. (2009) in a PLOS article on the topic, while the print article is well suited for human reading, it is highly problematic for machine reading, "thus inhibiting the development of services that can automatically link information between articles." Figure 3 shows semantic markup by key topics, with words in each topic (e.g. disease, habitat, organism) highlighted in a different colour, for a few paragraphs from an article appearing in *PLOS Neglected Tropical Diseases* (Reis et al. 2008). Figure 4 shows the markup displaying the supporting claims in the form of relevant quotations for one of the references. In the future such semantic linking and contextualizing of related articles could become a powerful tool to aid in the process of new knowledge generation, changing the research process in a fundamental way.

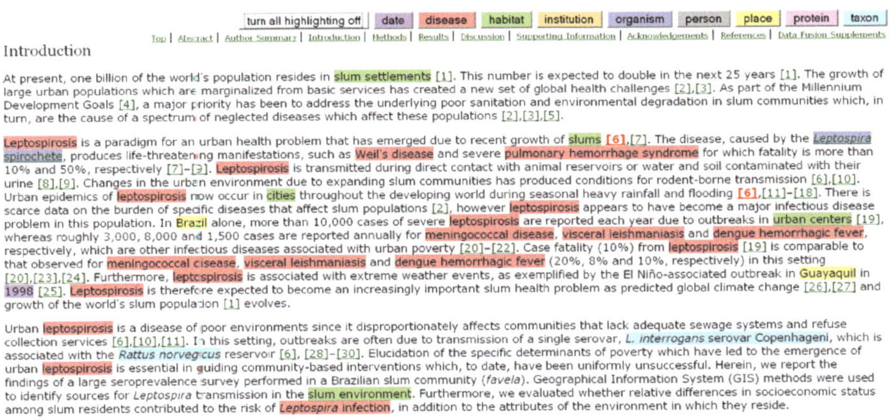
turn all highlighting off | date | disease | habitat | institution | organism | person | place | protein | taxon

Top | Abstract | Author Summary | Introduction | Methods | Results | Discussion | Supporting Information | Acknowledgements | References | Data Fusion Supplements

Introduction

At present, one billion of the world's population resides in slum settlements [1]. This number is expected to double in the next 25 years [1]. The growth of large urban populations which are marginalized from basic services has created a new set of global health challenges [2],[3]. As part of the Millennium Development Goals [4], a major priority has been to address the underlying poor sanitation and environmental degradation in slum communities which, in turn, are the cause of a spectrum of neglected diseases which affect these populations [2],[3],[5].

Leptospirosis is a paradigm for an urban health problem that has emerged due to recent growth of slums [6],[7]. The disease, caused by the Leptospira spirochete, produces life-threatening manifestations, such as Weil's disease and severe pulmonary hemorrhage syndrome for which fatality is more than 10% and 50%, respectively [7]–[9]. Leptospirosis is transmitted during direct contact with animal reservoirs or water and soil contaminated with their urine [8],[9]. Changes in the urban environment due to expanding slum communities has produced conditions for rodent-borne transmission [6],[10]. Urban epidemics of leptospirosis now occur in cities throughout the developing world during seasonal heavy rainfall and flooding [6],[11]–[18]. There is scarce data on the burden of specific diseases that affect slum populations [2], however leptospirosis appears to have become a major infectious disease problem in this population. In Brazil alone, more than 10,000 cases of severe leptospirosis are reported each year due to outbreaks in urban centers [19], whereas roughly 3,000, 8,000 and 1,500 cases are reported annually for meningococcal disease, visceral leishmaniasis and dengue hemorrhagic fever, respectively, which are other infectious diseases associated with urban poverty [20]–[22]. Case fatality (10%) from leptospirosis [19] is comparable to that observed for meningococcal disease, visceral leishmaniasis and dengue hemorrhagic fever (20%, 8% and 10%, respectively) in this setting [20],[23],[24]. Furthermore, leptospirosis is associated with extreme weather events, as exemplified by the El Niño-associated outbreak in Guayaquil in 1998 [25]. Leptospirosis is therefore expected to become an increasingly important slum health problem as predicted global climate change [26],[27] and growth of the world's slum population [1] evolves.

Urban leptospirosis is a disease of poor environments since it disproportionately affects communities that lack adequate sewage systems and refuse collection services [6],[10],[11]. In this setting, outbreaks are often due to transmission of a single serovar, *L. interrogans* serovar Copenhageni, which is associated with the *Rattus norvegicus* reservoir [6], [28]–[30]. Elucidation of the specific determinants of poverty which have led to the emergence of urban leptospirosis is essential in guiding community-based interventions which, to date, have been uniformly unsuccessful. Herein, we report the findings of a large seroprevalence survey performed in a Brazilian slum community (*favela*). Geographical Information System (GIS) methods were used to identify sources for *Leptospira* transmission in the slum environment. Furthermore, we evaluated whether relative differences in socioeconomic status among slum residents contributed to the risk of *Leptospira* infection, in addition to the attributes of the environment in which they reside.

Figure 3. Semantic markup by Shotton et al. (2009) of PLOS article by Reis et al. (2008)

It should be noted that, whatever their merits, the above variations on the conventions of the scientific article have not yet taken hold as conventional features in the PLOS journals or with any other publisher. The question is whether or not these merits outweigh the extra time and expense (by both authors and publishers) to implement such features. Worth keeping in mind is that the heading abstract was introduced in the 1920s, but did not become conventional practice until the 1950s (Gross, Harmon, and Reidy 2002: 176).

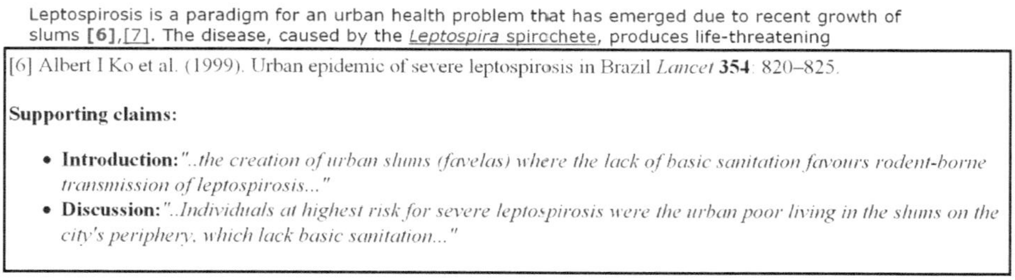

Figure 4. Semantic markup by Shotton et al. (2009) displaying supporting claims for reference in PLOS article by Reis et al. (2008)

5. Other innovations possible for the future scientific article

Of course, PLOS is not the only publisher or individual taking a stab at some sort of enhanced digital article. In around 2009, for example, one of the world's largest commercial publishers of scientific literature, Elsevier Publishing, unveiled a selection of prototypes for "The Article of the Future" (Aalbersberg et al. 2014). In the Elsevier article of the future, the screen is divided into three panes of different widths. The left is for navigation within the article, and most online articles now already have this feature. In this narrowest pane, the viewer sees the table of contents and thumbnails to all figures and tables. A click on a link in the left pane sends viewers to the corresponding location in the middle, where the article itself is displayed. In this middle pane, viewers can read or scan the whole article and save or print a PDF version; they can also watch any embedded videos and automatically convert tables into graphs. Finally, the right pane displays bibliographic information, keywords, and all kinds of supplemental information. For instance, if viewers click on an author's name in the middle pane, they will see in the right pane a biography and photograph. If viewers click on a link within the abstract, a video appears on the right and a podcast may be activated in which an author describes the article's contents. Click on a reference number in the article's text, and bibliographic information plus the abstract appear on the right along with a link to the referenced article. Click on an image in the text, and the image with caption appears in the right pane. Interactive images and videos also are found in the right pane. The data in some images are also tied into databases like the Protein Data Base, Google Maps, and ChemSpider. Click on an icon, and the viewer sees the information in the database on the right or middle pane. Another icon in the right pane grants readers access to all the gathered data for the article.

The Elsevier article of the future is not really an article of the future but an amalgamation of features already present in elite scientific journals. Whatever its merits, to date, nearly a decade after the unveiling of various prototypes, Elsevier's journals have applied it in a very limited fashion, especially as it applies to the right pane, the pane that really differentiates it from the typical current digital article.

A genuine scientific journal of the future was launched in 2006 by Moshe Pritsker with the first issue of the *Journal of Visualized Experiments*, which is still going strong today. He founded this journal to address the serious problem of repeating experiments based on the partial and sometimes confusing text in the methods section of journal articles, particularly in the biochemical and biomedical sciences, where repetition of experiments is fairly common for the purpose of verification or, more common, the pursuit of new research projects. Each article begins with a professionally produced video demonstrating some experimental procedure in the biological sciences, such as a method for obtaining RNA samples from human postmortem brain tissue. Accompanying that video are the typical components of a scientific methods article: a detailed abstract describing the method and its application; a step-by-step protocol for executing the method, often including additional visuals; representative results from applying the method combined with a discussion of their significance. In the past, scientists wanting to replicate a difficult experiment often had to spend weeks or months learning and perfecting a new technique or visiting the authors' laboratory for a demonstration. The goal of this journal is no less than greatly reducing this major inefficiency in scientific practice by creating a vast virtual repository of "visualized experiments" that feature video streams of scientists at work in the lab, intermixed with processes captured by video microscopy and microscopic structures rendered by 2D and 3D animation. One can easily imagine the day when videos of experimental methods are routinely a part of scientific articles in areas where replication based on print alone is not easily managed. For now, the *Journal of Visualized Experiments* has the field to itself.

A more recent innovative online-only scientific journal (2014) is *Royal Society Open Science*, which publishes a new genre called "registered reports" (Mehlenbacher 2019; Mehlenbacher and Mehlenbacher this volume). This peer-reviewed online-only journal issues such reports in two stages: first, the methodologies for research projects "before data collection commences," then the full article with results and discussion after data collection and analysis has ended. Unlike typical journal articles, *Royal Society Open Science* publishes the final registered report whatever the final outcome, even if the results do not support the main thesis. The goal of these registered reports is to diminish "publication bias while minimizing, as much as possible, various forms of researcher bias." In other words, unlike traditional journals, even if the research project fails in its main purpose, the journal publishes the negative findings, which can help other researchers to avoid dead ends. Importantly, these reports differ from print scientific articles in that they center "the reporting of science in serious methodological debates" (Mehlenbacher 2019: 38).

Another possible major shift in the transition from the print to online genre concerns the digital information repository. As noted by Gross and myself in *The Internet Revolution of the Sciences and Humanities* (Gross, and Harmon 2016: 91), digital information repositories are sporadically springing up online whose primary purpose is to provide resources for scholars, store data for scientific research, create new knowledge through voluntary participation in a research project, and codify existing knowledge. Examples in the sciences include the Protein Data Bank, a database with structural data and visualization of over 100,000 large biomolecules; The Encyclopedia of Life, whose goal is to store all research information on all the earth's plants and animals; and ChemSpider, a database with chemical information and visuals on over 63 million molecules. For the individual entries – whether biomolecule, chemical, plant, or animal – these information repositories are also including linked lists of published research papers relevant to the topic. The digital scientific article is thus finding a new ecosystem in which it might thrive in a way not possible before. The day might arrive when all scientific articles find their way into such knowledge repositories, a more suitable resting place than a journal issue filled with loosely connected articles on some broad topic.

Finally, a relatively small but possibly important change to the article of the future could make it even more open than just open for all to read: inclusion of links to the reports by the peer reviewers and author response (Gross and Harmon 2016: 135–144; see also Breeze this volume). Started in the 17th century, though not a convention until the early 20th, peer review has been mostly conducted through the postal system and been open only to the authors, peer reviewers, and editors involved, with the names of the reviewers withheld from the authors. With the Internet, the peer review process can be easily opened to all by adding links to the communications involved, with or without the reviewers being named. The major communicative advantage is that such links would give readers access to the critical commentary that shaped the decision to publish, gaining an expert's perspective on the content and the thinking behind the judgment of its worthiness. To date, while open peer review has met with limited success, a few important journals are testing this innovation to different degrees, including *Royal Society Open Science*, *Nature Communication*, and fourteen journals published by the European Geosciences Union. The peer review report may one day add yet another genre to the rich blend of genres that is part of the online scientific article.

6. Concluding remarks

The good news for the genre of the scientific article is that its future seems as bright as ever. The Internet is not going to make the scientific journal obsolete any time soon. Electronic journals are proliferating, as evidenced by not only the success of PLOS but the many spinoff journals from prestigious journals like *Science* and *Nature.* After the relative stasis of the scientific journal article in the 19th and 20th century, moreover, proliferation of communicative innovations seems to be a norm in the elite journals. Given these innovations, it seems reasonable to project that the online scientific article as a genre will one day conform to the following conventions:

- The verbal text is highly interactive in that links send readers to scientific articles cited in the original, any relevant news coverage and blogs, comments and tweets by readers, related scientific articles selected on the basis of a screening algorithm, peer review reports and author responses (see also Breeze this volume), and any articles that cite the original. The "text" is thus a field of links competing for the readers' attention.
- Many of the visuals are interactive and dynamic, and are woven into the verbal text rather than buried in Supplemental Information.
- The experimental section includes the usual description of materials and methods, plus a video of the authors performing the experiment if likely that others may want to reproduce it (see Rowley-Jolivet and Carter-Thomas this volume).
- All the data gathered from applying the experimental method are accessible and downloadable into a spreadsheet through a link.
- Semantic linking makes the scientific literature itself analyzable in a way never before possible.
- The articles themselves are incorporated into large bodies of knowledge in the form of digital repositories such as the Protein Data Bank and Encyclopedia of Life.

In closing, it is also worth briefly mentioning some of the possible problems with the "article of the future." The ease of uploading scientific articles to the web has made control over the quality of the scientific literature more challenging. Given the ease of online publication, it seems safe to infer that the overall technical quality of the scientific literature is as low as it has ever been. Anyone can post almost anything somewhere, including articles with fabricated data or other major failings, making it harder than ever to separate the wheat from the chaff in a literature search. For example, predatory journals are now charging authors to publish their papers, where these papers undergo no peer or editorial review whatsoever (Beal

2018). It also seems possible that, in an Amazon.com-like effect, a few prominent publishers could monopolize scientific publishing. Also problematic, at least for highly newsworthy or controversial articles, has been the distortion of the science from scientific articles that sometimes occurs in blogs, tweets, and even mainstream newspapers (Fahnestock 2016). No one fully appreciated the possible damaging effects of social media until recent developments in the political arena; we do not yet know whether a similar communicative problem might one day bedevil the scientific literature.

References

Aalbersberg, IJsbrand J., Sophia Atzeni, Hylke Koers, Beate Specker, and Elena Zudilova-Seinstra. 2014. "Bringing Digital Science Deep Inside the Scientific Article: the Elsevier Article of the Future Project." *LIBER Quarterly* 23(4): 274–299. https://doi.org/10.18352/lq.8446

Atkinson, Dwight. 1999. *Scientific Discourse in Sociohistorical Context: The Philosophical Transactions of the Royal Society of London*, 1675. Mahwah, NJ: Erlbaum.

Banks, David. 2008. *The Development of Scientific Writing: Linguistic Features and Historical Context*. Oakville, CT: Equinox Publishing.

Barber, Charles. 1962. "Some Measurable Characteristics of Modern Scientific Prose." In *Contributions to English Syntax and Philology*, ed. by John Swales, 1–23. Stockholm: Almquist & Wiksell.

Bazerman, Charles. 1988. *Shaping Written Knowledge: The Genre and Activity of the Experimental Article in Science*. Madison, WI: University of Wisconsin Press.

Beal, Jeffrey. 2018. "Beal's List of Predatory Journals and Publishers." Last accessed July 4, 2018. https://beallslist.weebly.com

Biber, Douglas, and James K. Jones. 2005. "Merging Corpus Linguistic and Discourse Analytic Research Goals: Discourse Units in Biology Research Articles." *Corpus Linguistics and Linguistic Theory* 1–2: 151–182.

Buehl, Jonathan. 2016. "Revolution or Evolution? Casing the Impact of Digital Media on the Rhetoric of Science." In *Science and the Internet: Communicating Knowledge in a Digital Age*, ed. by Alan G. Gross and Jonathan Buehl, 1–9. Amityville, NY: Baywood Publishing.

Casper, Christian F. 2016. "The Online Research Article and the Ecological Basis of New Digital Genres." In *Science and the Internet: Communicating Knowledge in a Digital Age*, ed. by Alan G. Gross and Jonathan Buehl, 77–98. Amityville, NY: Baywood Publishing.

Fahnestock, Jeanne. 2016. "Controversies on the Web: The Case of Adult Human Neurogenesis." In *Science and the Internet: Communicating Knowledge in a Digital Age*, ed. by Alan G. Gross and Jonathan Buehl, 117–141. Amityville, NY: Baywood Publishing.

Giltrow, Janet. 2011. "'Curious Gentlemen': The Hudson's Bay Company and the Royal Society, Business and Science in the Eighteen Century." In *Writing in Knowledge Societies*, ed. by Doreen Starke-Meyerring, Anthony Paré, Natasha Artemeva, Miriam Horne, and Larissa Yousoubova, 53–74. West Lafayette: Parlor Press.

Giltrow, Janet, and Dieter Stein (eds.). 2009. *Genres in the Internet: Issues in the Theory of Genre*. Amsterdam, Netherlands: John Benjamins. https://doi.org/10.1075/pbns.188

Gross, Alan. 2016. "Revolution or Evolution? Casing the Impact of Digital Media on the Rhetoric of Science." In *Science and the Internet: Communicating Knowledge in a Digital Age*, ed. by Alan G. Gross and Jonathan Buehl, 59–76. Amityville, NY: Baywood Publishing.

Gross, Alan G., and Joseph E. Harmon. 2013. *Science from Sight to Insight: How Scientists Illustrate Meaning*. Chicago: University of Chicago Press. https://doi.org/10.7208/chicago/9780226068343.001.0001

Gross, Alan G., and Joseph E. Harmon. 2016. *The Internet Revolution in the Sciences and Humanities*. New York: Oxford University Press.

Gross, Alan G., Joseph E. Harmon, and Michael Reidy. 2002. *Communicating Science: The Scientific Article from the 17th Century to the Present*. New York: Oxford University Press.

Gross, Alan G., and Jonathan Buehl (eds.). 2016. *Science and the Internet: Communicating Knowledge in a Digital Age*. Amityville, NY: Baywood Publishing. https://doi.org/10.4324/9781315231099

Harmon, Joseph. 2016. "The Scientific Journal: Making it New?" In *Science and the Internet: Communicating Knowledge in a Digital Age*, ed. by Alan G. Gross, and Jonathan Buehl, 33–58. Amityville, NY: Baywood Publishing.

Hobaiter, Catherine, Timothée Poisot, Klaus Zuberbühler, William Hoppitt, and Thibaud Gruber. 2014. "Social Network Analysis Shows Direct Evidence for Social Transmission of Tool Use in Wild Chimpanzees." *PLOS Biology* 12(9): e1001960. https://doi.org/10.1371/journal.pbio.1001960

Jones, Benjamin F., Stefan Wuchty, and Brian Uzzi. 2008. "Multi-university Research Teams: Shifting Impact, Geography, and Stratification in Science." *Science* 322: 1259–1262. https://doi.org/10.1126/science.1158357

Karlberg, Tobias, Susanne van den Berg, Martin Hammarström, Johanna Sagemark, Ida Johansson, Lovisa Holmberg-Schiavone, and Herwig Schüler. 2009. "Crystal Structure of the ATPase Domain of the Human AAA$^+$ Protein Paraplegin/SPG7." *PLOS One* 4(10): e6975. https://doi.org/10.1371/journal.pone.0006975. See also video at http://global.oup.com/us/companion.websites/9780190465933/videos/ch2/video2.4

Klinkhamer, Ada J., D. Ray Wilhite, Matt A. White, and Stephen Wroe. 2017. "Digital Dissection and Three-Dimensional Interactive Models of Limb Musculature in the Australian Estuarine Crocodile (*Crocodylus porosus*)." *PLOS One* 12(4): e0175079. https://doi.org/10.1371/journal.pone.0175079

Mehlenbacher, Ashley R. 2019. "Registered Reports: Genre Evolution and the Research Article." *Written Communication* 36 (1): 38–67.

Miller, Carolyn R. 1984. "Genre as Social Action." *Quarterly Journal of Speech* 70: 151–167. https://doi.org/10.1080/00335638409383686

Owen, John M. 2007. *The Scientific Article in the Age of Digitization*. The Netherlands: Springer.

Reis, Renato B., Guilherme S. Ribeiro, Ridalva D. M. Felzemburgh, Francisco S. Santana, Sharif Mohr, Astrid X. T. O. Melendez, Adriano Queiroz, Andréia C. Santos, Romy R. Ravines, Wagner S. Tassinari, Marília S. Carvalho, Mitermayer G. Reis, and Albert I. Ko. 2008. "Impact of Environment and Social Gradient on Leptospira Infection in Urban Slums." *PLOS Neglected Tropical Diseases* 2(4): e228. https://doi.org/10.1371/journal.pntd.0000228

Shapin, Steven, and Simon Schaffer. 1985. *Leviathan and the Air-Pump: Hobbes, Boyle, and the Experimental Life*. Princeton: Princeton University Press.

Shotton, David, Katie Portwin, Graham Klyne, and Alistair Miles. 2009. "Adventures in Semantic Publishing: Exemplar Semantic Enhancements of a Research Article." *PLOS Computational Biology* 5(4): e1000361. https://doi.org/10.1371/journal.pcbi.1000361

Sidler, Michelle. 2016. "The Chemistry Liveblogging Event: The Web Refigures Peer Review." In *Science and the Internet: Communicating Knowledge in a Digital Age*, ed. by Alan G. Gross, and Jonathan Buehl, 99–116. Amityville, NY: Baywood Publishing.

Swales, John M. 1990. *Genre Analysis: English in Academic and Research Settings*. Cambridge: Cambridge University Press.

Weinberger, David. 2012. *Too Big to Know: Rethinking Knowledge Now That the Facts Aren't the Facts, Experts Are Everywhere, and the Smartest Person in the Room Is the Room*. New York: Basic Books.

Wickman, Chad. 2016. "Learning to "Share your Science": The Open Notebook as Textual Object and Dynamic Rhetorical Space." In *Science and the Internet: Communicating Knowledge in a Digital Age*, ed. by Alan G. Gross, and Jonathan Buehl, 11–22. London: Routledge.

Wuchty, Stefan, Benjamins F. Jones, and Brian Uzzi. 2007. "The Increasing Dominance of Teams in Production of Knowledge." *Science* 316: 1036–1039. https://doi.org/10.1126/science.1136099

CHAPTER 3

The case of the scientific research article and lessons concerning genre change online

Ashley Rose Mehlenbacher and Brad Mehlenbacher
University of Waterloo

Although the apparatus surrounding the scientific research article (SRA) has evolved to include hyperlinks, comments, and supplementary data in online repositories, the genre itself seems stubbornly committed to its form. In itself, the lack of change is an important and notable case for genre scholarship. Recently, however, the replication crisis in the psychological and life sciences has created an exigence for genre change. The "Registered Report" is an emerging genre of SRA that responds to issues raised by the replication crisis that we will examine in this chapter. Rather than characterizing the genre in linguistic terms alone, we advance a rhetorical inquiry into the recurrent rhetorical situation to which this permutation on the scientific research article responds.

Keywords: genre change, genre evolution, research genres, scientific research article, trans-scientific genres

1. Introduction

It seems that, in many aspects of daily life, genres are proliferating at a blistering pace. Not only do genres seem to migrate online from earlier forms, but they also seem to be borne digital, or at least to emerge as a bricolage of familiar forms (see also Rowley-Jolivet and Carter-Thomas this volume). Further complicating matters for genre theorists, the rapid evolution of online forms continues to challenge genre classification as such forms resist stabilization. Genre theory has grappled with how the web influences genre evolution and change for decades (McNeill 2003; Miller and Shepherd 2004, 2009; Herring et al. 2005; Giltrow and Stein 2009; Garzone 2012; Miller and Kelly 2017). Although these are important issues for genre theorists, here we draw attention to something of a curmudgeonly genre, a genre that seemingly resists the tide of genre change online: the scientific research article. In this chapter, we draw attention – in the midst of

https://doi.org/10.1075/pbns.308.03meh
© 2019 John Benjamins Publishing Company

significant genre emergence, evolution, speciation, and change for online genres of science communication – to why a genre that we might comfortably state is "stabilized-for-now" (Schryer 1993) is important to discussions of genre change. We investigate the scientific research article through the lens of genre in an effort to better understand how the research article has helped shape our understanding of genre stabilization and change, and where the scientific research article may continue to carve out compelling directions for genre researchers interested in new media environments.

Although the research article itself has remained a stabilized genre, the broader environment that the genre inhabits has seen considerable evolution and change (see also Harmon, and Maier and Engberg this volume); pre-publication repositories, associated datasets, multiple graphical representations of the data, blog networks committed to sharing research article findings, post-publication review, the *Retraction Watch* website, Twitter, etc., all change the ways in which research articles are accessed and their findings shared. In its stability in this online ecology, the scientific research article demands that genre scholars return to it anew. While it is important to consider the online media ecologies in which the scientific research article genre is situated, this offers only a partial story of the genre's function. Instead, we must also examine the rhetorical context in which the genre circulates, understanding the situations, exigencies, and norms within which the genre is invoked. For the scientific research article, a highly regularized community of practice shapes the norms of the genre; or, we might say, communities of practice, if we attend in a more nuanced manner to specific fields. Given the constraints of the genre imposed by the community of practice, the community of genre users, the seemingly negligible influence those online media affordances appear to have had on the genre of the research article itself is perhaps not surprising.

Where a highly stabilized genre leaves us to consider, as the collected chapters in this edited volume encourage us to do, the relationship between "old" and "new" genres of science communication on the Internet may, at first, seem something of a dull provocation – or, if not dull, at least uninspired. What we argue, however, is that the seeming stabilization of this genre is perhaps a matter of where we focus our attention. Although it seems as though the scientific research article is largely still organized following the Introduction, Methods, Results, and Discussion (or, IMRAD) model, some permutations have been noted, as the IRADM model (Berkenkotter and Huckin 1995; Miller and Fahnestock 2013), where the methods section has been placed at the conclusion of the article.

We are motivated to return to the question of rhetorical situation by an interesting case in the life and psychological sciences. Recently, an old question of how well research in these fields can be replicated has been given new life, notably by a large-scale replication study that produced rather unsettling results, namely only a 39% replication rate, and the popular press coverage that subsequently followed. Resulting in the current "replication crisis," scientists have been debating what can be done to produce more replicable research. One response has been the advent of "Registered Reports," which describe a new model of journal article involving two-stage peer review (Chambers 2013; Chambers et al. 2015). Registered Reports are a direct response to the contemporary replication crisis, and their generation and implementation is made possible through online technologies. In this model, a research protocol consisting of the introduction and methods section of a paper is peer reviewed, and later, provided the first round of review was successful, the full article inclusive of a results and discussion sections, is again subject to peer review. Although these reports seemingly have a neat exigence and emerging form we can chart, the situation from which Registered Reports emerge, and their broader significance to genre theory merits continued attention. Registered Reports is a case that demonstrates how genre evolution is married with not only technological change, but also fundamentally by the community of genre users and their larger embedded social and institutional contexts.

As we will argue, too, Registered Reports mark an important rupture in the scientific research article genre. It is perhaps worth briefly aligning ourselves with a particular notion of rhetorical genre. By genre we follow Miller's (1984: 159) definition of genre as "typified rhetorical actions based in recurrent situations." Miller further explains that we should understand a genre as "complete" or distinct in "that they are circumscribed by a relatively complete shift in rhetorical situation" and, thus, how we recognize genres follow from "our determination of the typified rhetorical situation" (Miller 1984: 159). Indeed, following a Milleresque approach, we see the scientific research article genre as not only particular invocations and their particular structural, stylistic, or linguistic forms, but rather as a combination of those features in typified forms, along with the crucial stance of those typified forms as responses to recurrent rhetorical situations. It is the latter, the recurrent rhetorical situation, where we uncover the emergence of this proto-genre of scientific research articles, the Registered Report. Indeed, as we will show, the form of the genre may lead us to believe no genre change has occurred, but attention to the structural features of the articles alone would miss the crucial rhetorical functions that operate in the genre.

In the case of Registered Reports, the genre change we chart is one that marks a rather fundamental shift in thinking about the purpose of the research article genre, a kind of recovery of older conceptions of the genre's functions. Thus, matters of old and new genres are complicating progressive narratives about the genre's function, and instead encourage us to understand the reflexive and recursive nature of genres. Put more straightforwardly, the genre change we find in Registered Reports gestures toward important tensions in the scientific community, and because we believe genres are important elements in constructing and reflecting community norms and their epistemic commitments, we believe a rhetorical account attendant to the exigence invoking genres is an important approach to accounting for the change in genre, from old to new. Indeed, Registered Reports are a compelling case for genre scholars interested in new media environments because they allow the relationship between a highly-stabilized, powerful genre and the community of genre users, antecedent genres, genre systems and ecologies, and new media to be examined in a single case.

2. Evolution of the scientific research article genre and the evolution of online genres of science communication

In genre theory, the scientific research article has been cast as the "primary genre in science" (Miller and Fahnestock 2013). Early work by Bazerman (1988) set out to detail the evolution of the experimental report. Bazerman provides foundational work for rhetoricians and genre theorists interested in the stylistic and formal particulars of how experiments were reported along with aspects of cultural and institutional changes occurring as the enterprise of science evolved. Berkenkotter and Huckin (1993: 478) provided five principles for theoretically framing genres, including their "dynamism" (the way they both respond to recurrent situations and stabilize experience), "situatedness" (how they develop and embed our engagement in communicative activities), "form and content" (shaping content for particular purposes and situations), "duality of structure" (genres form and reproduce social structures), and "community ownership" (conventions of genres form discourse communities' norms and ways of thinking). These principles are helpful for thinking about the constraints of a genre, how a genre might evolve over time, and where lines of influence for genre change might be traced. We also wish to note that genre dynamism, situatedness, and community ownership are crucial to the questions of genre change we wish to examine in our case of the scientific research article. These features, in addition to questions of form and content and the duality of structure, are integral to a rhetorical conception of genre, and indeed, genre change.

As we noted, for the scientific research article, genre change has been slow. Although the rise of the web has dramatically changed many modes of communication, it seems the research article itself has been a hold out, while a system of coordinating genres and technologies have continued to evolve around it. Owen found that very little change had occurred to the scientific research article when he published in 2007. Subsequent studies found that there had been unexpectedly little change in the research article, although, again, certainly features have developed around the article itself (Kittle Autry 2013; Casper 2016; Gross and Harmon

2016; Harmon 2016 this volume). This is not to suggest that the scientific research article resists change completely, especially given that genres are dynamic and develop over time in response to their recurrent and evolving use. Thus, Li and Ge (2009) found that articles have in fact changed over time, in the case of their study; moreover, many of these changes could be captured by Swales' (1990) move analysis, and later work by Nwogu (1997), where certain moves became optional versus obligatory and vice versa over time (Li and Ge 2009: 96–99). Changes in the structure of argument are important features of genre change, and these suggest that, although appearing slow to change, change within the genres has occurred.

By way of example, we might examine the Public Library of Science journal platform, which is an interesting case of a borne digital scientific journal publishing enterprise (see also Harmon this volume). Further, the platform has had significant success and, thus, may have more flexibility in their experimentation with the scientific journal article genre; or, at least, the apparatus surrounding each article. A number of different features and genres appear embedded in or linked through the research article genre. Formal features of a research article appear prominently, with the title and authors opening the article, and a menu linking to various sections (following the IMRAD logic of an Introduction, Methods, Results, and Discussion arrangement of sections, and also with additional sections for "Supporting Information"). Around the article various informational widgets are provided, noting the views, shares, saves, and a citation count for any given article. Options to download, print, or share the article are provided, and tags or keywords provide further contextualization and mapping. Readers may comment on articles, and media coverage of the article is linked as well. Within what we might call the article itself, hyperlinks to references, tools for more closely examining images, annotation options, and linking to tables and figures throughout provide new means to view, explore, and navigate parts of the article but do not fundamentally alter the IMRAD article structure. Articles appear to follow similar stylistic and formal aspects that the scientific research article has been defined by for decades. Although the features of the article remain intact, new modes for engaging with these articles certainly emerge, and Gross and Harmon (2016: 50) detail how these different modes for engagement shape the evolution of the scholarly journal. Their conclusion, however, reaffirms that the article itself remains highly stabilized:

> Readers not only read but interact with the digital article. The web does not redefine the essence of the scientific article; it has always been a node in a network of knowledge. Rather, the web permits the exploration and exploitation of possibilities inherent in the structure of scientific communication. The article is what it always was, only more intensely, more completely.

But why is this so? One rationale might be that, although the web and modern computational tools have brought about changes to how science is conducted, those changes have been in the service of a broader intellectual framework that directs how scientists become enculturated to and engage in knowledge-making activities. Broadly, this might be described as the scientific method or the Mertonian values ascribed to science, but very specifically we might look at the institutional contexts that drive scientific research. Researchers in academic institutions conduct a good deal of basic research where the purpose of their work is not necessarily to provide immediate research outcomes for broad publics. Indeed, researchers normally explicitly delineate activities related to their research from activities related to extension and public engagement as part of their academic review process. Importantly, scientific research is incremental, and when authors write in the scientific research article genre, they likely have in mind their immediate audience of peers who may take this small advancement and build on it within their own research. Although a grossly generalized example, we want to stress that critiques of research articles as not serving broad audience demands are often overly simplistic. Research articles are not intended to rhetorically function in this way. Instead, in their most fundamental form, given their length, details, and the kind of content they are meant to share, research articles are specifically designed for highly specialized scientific audiences. Certainly, critiques of this vein still help to advance the scientific research article by suggesting approaches that make its contents accessible to interested researchers or expert publics, to a global community of researchers, and to science journalists and others who serve to accommodate (Fahnestock 1986) scientific research article contents to broader audiences. As scientific research articles are accommodated (Fahnestock 1986) to broader, public audiences, the scientific research article continues to serve distinct social actions. Reporting on articles shows publics the various ways in which science enriches our lives, from breakthroughs related to health, or those, as Fahnestock describes them, that appeal to our sense of wonder. But the mystery and wonder that makes science so appealing to many is not universal, and some worry that, with the rise of anti-science messaging, the so-called replication crisis might be used to further undermine areas of research already distrusted by publics (see an example of this in the popular press: Schulson 2018). With multiple rhetorical purposes and audiences, the research article remains not only a complex genre of scientific research, but a highly important one. Further, the research article is embedded within a broader social negotiation of science in society. It is perhaps unsurprising, then, that a reconfiguration of the research article is one of the responses to the purported replication crisis.

However, genre theory helps us untangle how the scientific article itself might be embedded within a system of genres that, in concert, work to respond to the multiple rhetorical demands made of the research article genre itself. In this way, we can see that the scientific research article in its current form appears to respond to its primary rhetorical exigence: to share research *findings* with experts. We only need to examine what happens when scientific research findings are communicated to publics prematurely or prior to verification and/or replication. In environmental science, this becomes particularly problematic when publics with deep-seeded emotional attachments to particular beliefs and perspectives review scientific research that they have limited training interpreting (Schmidt 2009). Issues around climate change, for example, generate what Schmidt (2009) describes as "mental shortcuts" that invite misunderstanding, but they are more than that – publics engage with scientific research as discourse novices who rarely engage the research article genre and its long-term use, extension, and verification processes. Similar to the scientific journal article, many scientific studies are complex, nuanced, and difficult to explain to non-scientific audiences, and for good reason: science relies on incremental and painstakingly careful reporting and review.

Yet, online scientific articles are immersed in a number of conversations, both scholarly and more public, about how the communication of science ought to catch up with social and technological changes. Conversations about Open Access publishing suggest that research should be made freely available to anyone who wishes to read it, rather than being cloistered behind a paywall that renders that research accessible only to a privileged few who have institutionally-supplied access to expensive journals. Open Access to data is likewise a concern for researchers, and is increasingly dealt with by major granting agencies requiring data be shared. Access to data is important, as it is one means to vet what is reported in an article, and may be provided as a supplemental to the article, or otherwise located in a repository where other researchers may access and use it. Further issues regarding peer review also implicate the scientific research article. Proposals for pre- or post-publication review change the model of how articles might be published. Others recommend more transparent forms of review, where perhaps reviewers identify themselves or reviews are made publicly available (see also Breeze this volume). Criteria for reviews themselves are under scrutiny by some journals, such as the Public Library of Science's flagship journal *PLOS ONE* (Kelly and Kittle Autry 2013). It seems the scientific research article might single-handedly be holding the entire enterprise back with its old-fashioned form – or so articles such as *The Atlantic's* "The Scientific Paper is Obsolete" might suggest (Somers 2018).

"Obsolete" is an interesting adjective to apply to a genre that has been evolving for hundreds of years, and continues to do so today, suggesting the research article might simply be too old fashioned for the rhetorical needs of today's scientists. Somers (2018) writes that the "basic means of communicating scientific results hasn't changed for 400 years," but of course this is an imprecise measure of what has changed. Genre theorists including Bazerman (1988), Gross, Harmon, and

Reidy (2002), Swales (2004), and Berkenkotter and Huckin (1995), among others, have demonstrated that the scientific research article has indeed evolved significantly over this period, responding to changing rhetorical demands. Rhetorical awareness renders moot the question with which Somers (2018) follows up this provocative statement: "What would you get if you designed the scientific paper from scratch today?" "From scratch" suggests we might imagine rhetorical ends divorced from the *topoi* of science, the antecedent genres and the interconnected genre systems (e.g. abstracts, conference talks, and peer reviews) that have developed incrementally over time and that the scientific research article participates in and relies upon. Much of Somers' article is focused on technical solutions, noting the uptick in how immersed technical processes are in the course of scientific research. Resituating the lament in a rhetorical framework, we might instead ask why it is that the lack of changes to scientific research articles draws such ire while simultaneously remaining so highly stabilized. This might seem a rather rudimentary argument to those rhetorical, genre, and communication scholars who understand the varying needs and complexities of different audiences, but the calls to make research articles accessible, jargon-free, and easily understandable suggest indeed that the lesson bears repeating (Yong 2010; Ball 2017; Somers 2018). And, further to this point, that genre researchers have tools to explain not only why this is important, but also tools to help demonstrate how to assess audience and find more tools in one's rhetorical toolbox to respond effectively to the given exigence.

Calls to reinvent the scientific research article, however they may misread the primary rhetorical purpose of the article, importantly mark broader changes to how scientific knowledge circulates or is expected to circulate, in a contemporary, web-based society. When Ashley R. Mehlenbacher and Carolyn R. Miller (2017) began to examine how web-based genres were employed following the 2011 nuclear disaster at Fukushima Daiichi, they argued that a new phenomenon of genres that operate outside of the traditional scientific genres were being used. These genres were not what we might call research process genres, following Swales (1990, 2004), such as the scientific research article, but they also were not popularizations, as one might conventionally understand. Rather, Mehlenbacher and Miller (2017) argue that these are para-scientific genres (expanding Kaplan and Radin's 2011 notion of para-scientific) that operate alongside traditional genres of science communication, relying on certain markers of epistemic authority common among scientific genres, but that also work to communicate with a broader public than research scientists. That is to say, these genres blur the boundaries between what are traditionally conceived of as professional genres of science communication and popular genres of science communication.

Consider, for example, science blogs. Science blogs include a range of genres to share scientific methods, findings, or even disciplinary controversies with researchers, teachers, citizen scientists, and broader publics. Although science blogs might describe findings from a research article – such is the model for many blog posts on the Public Library of Science blog network, which is connected with the broader Public Library of Science effort – these blogs do not serve the same function as an article. They come to support efforts to communicate the findings of scientific research more broadly, and they in part respond to current calls for forms that allow broader audiences to read accessible descriptions of scientific research. Likewise, Twitter and other social media platforms allow such content to be distributed to different networks of readers, commanding attention to research in a way that is quite accessible as an introduction, and also one that provides information on how to access the original article itself (see also Orpin this volume; Reid and Anson this volume; Smart and Falconer this volume). Instead of demanding that a genre be reinvented or evolve to meet multiple, often times conflicting, rhetorical demands, these new, online genres expand the range of genres working in concert to communicate scientific research to multiple audiences (see also Mehlenbacher 2019a). In these genres we find another rationale for why the scientific research article may not be evolving: scientists and those communicating science to broader publics understand the multiple rhetorical needs of their heterogeneous audiences and they are responding by exploring a wide range of online rhetorical activities. Again, this suggests that the original research article continues to serve a primary function within expert communities to economically communicate complex research, methods, and findings without losing fidelity.

While these online genres are beginning to respond to various needs, and the scientific article seemingly continues to serve many of its primary rhetorical functions, some scientists believe the research article itself is at the heart of a widespread scientific crisis. This crisis, however, is not one of communicating to an audience for which the genre was not designed. Rather, the crisis is much deeper, at the heart of the scientific enterprise itself. The aforementioned "reproducibility crisis," or the "replication crisis," marks a critical rupture to business as usual for scientific publishing. Foundational to the scientific method, reproducing research independent from the researchers who originally conducted a study helps demonstrate that the research methods and approach are sound and free from bias. However, in the current publishing model, there is little incentive for replication studies, and, further to that, considerable incentive to publish highly original, impactful research. The scientific research article itself, then, becomes a focus of the replication crisis. With this considerable attention to the failures of current publishing models inclusive of the scientific research article, a new genre emerges in response to this exigence, marking a notable shift in the rhetorical situation. We chart this movement in the next section.

3. The replication crisis: An exigence for genre change in the scientific research article

In 2015, the Open Science Collaboration conducted a ground-breaking empirical study to assess how often studies could be reproduced, and published the results of their effort in *Science*, in an article entitled "Estimating the reproducibility of psychological science." It provided concerning results. The authors attempted to replicate 100 studies and argued they were only able to replicate 39% (Nosek et al. 2015). As noted, problems of replication date back several decades, but this 2015 study proved a catalyst for rethinking questions of replication problems today. Even the name of the collaboration gives us some indication about important movements influenced by the web. Open science, of course, could be traced back pre-web, in terms of cultures of sharing through journals, for example. But the conception today of open science, tied to open databases, open access articles, and so on, are commitments that can be afforded by the web. Indeed, more radically, commitments such as sharing large datasets through public domain licenses and freely sharing published scholarship globally are made possible online. While the web did not set in motion the replication crisis, it certainly afforded new responses. One such response is the emerging genre of the Registered Report. Different researchers have posed registered studies as a solution at different times, but the current imagining is embedded within the framework of open science. Slowly this approach is gaining traction across fields of psychological, life sciences, and beyond. According to the Center for Open Science, Registered Reports are now featured in more than 190 journals.

A Registered Report takes a highly similar *form* to the traditional research article, complete with abstract, references, as well as the features to link, share, cite, and so on, in digital versions. Registered Reports appear to follow the IMRAD logic of research articles in the sciences, first Introducing the research, then providing Methods, Results, and Discussion. But this assessment of the Registered Report at the level of form misses fundamental changes brought about by this emerging genre. Instead, we must look to the rhetorical situation that generates an exigency to which this genre responds. The exigence to which Registered Reports respond is, in part, the ongoing replication crisis, renewed with urgency following the Open Science Collaboration's 2015 study. The crisis itself is, of course, composed of a variety of problems in research, including what are called questionable research practices such as the infamous p-hacking or the lack of null results reported in research, which is to say the preoccupation in publishing "novel" and "original" findings rather than uncertain or unremarkable findings. It is this exigence that marks a relatively new rhetorical situation, one that calls for a different genre than the traditional scientific research article. The crisis, as exigence, marks a broader rhetorical situation in which science must be conducted as a replicable practice, to thus ensure credibility and legitimacy of the enterprise. While in the past problems of replication were noted, they were perhaps less notable, and thus seemingly did not demand genre innovation. Research articles, as we have established, have largely maintained many of the typifications that we can say constitute the research article genre.

Further, we can trace the change to the genre through the writing and reviewing practices that ultimately shape genre participation. Where Registered Reports alter the genre of the scientific research article is found in how they change the peer review process (for changes in the peer review process, see Breeze this volume). In the Registered Report model, a research study is first proposed – we can call this stage 1 – detailing the theoretical framework and methods to be used (the I and M of the IMRAD model), and this proposal undergoes a round of peer review. At this stage, the quality of the methods and their situating become the focus of the review. Provided a study can be justified, it should pass through this first round of peer review. At this point, researchers begin the study, from data collection to analysis, and then report their findings in a full article (adding the R and D, resulting in the common IMRAD structure in the final article text). The full article is sent out once again for peer review – we can call this stage 2. During this second round of review, the reviewers reassess the methods to ensure they were followed as described in the proposed research. At this stage, the results and explanatory framework are also assessed.

In Mehlenbacher (2019b), a case analysis of Registered Reports published in the journal *Royal Society Open Science* (RSOS), and a small corpus of 32 stage 1 protocols and 77 stage 2 reports from the Open Science Framework's Zotero repository were examined for evidence of genre change. To understand how the rhetorical situation for the research article had changed, *RSOS* provides not simply a journal to explore published articles, but reports the entire review process, including reviewer reports and correspondence between the editors and authors. Crucially, the correspondences reveal an abiding commitment on behalf of the editorial staff to reject reviewer assessment of originality. That is to say, that the foundational criterion of originality was not a criterion deemed appropriate for Registered Report articles. Such a significant shift in the assessment of scientific research articles marks an important change in the rhetorical situation to which such articles respond. Another important finding of Mehlenbacher (2019b) was that the IMRAD model common among scientific research articles continues to be the general structure for Registered Report-based articles. Again, the structure of a published Registered Report article might very well appear indistinguishable from a regular scientific research article. However, the shift in the rhetorical situation that marks Registered Reports means the genre function is quite distinct.

Indeed, it would be those methodological concerns such as p-hacking and other questionable research practices that would be absent from the Registered Report articles. These practices, as the crisis itself has taught us, are not easily identified in the text itself, but rather are revealed in the scientific method. By their nature, these practices are not directly reported, and thus make the assessment of research without replication difficult. Registered Reports are an alternative in that one would be able to trace back the methodological design, data collection, and analysis plans to a degree uncommon among traditional scientific research articles.

But there are textual clues to the distinctions in the genres, as reported in Mehlenbacher (2019b). Final Registered Reports may include a note about their status and even reference to the stage 1 protocols. In those stage 1 protocols the absence of discussion and methods, as one would expect, renders a notable textual distinction. Further, methods are highly detailed, sometimes including bulleted lists with extensive detail about procedures and materials to be used. As the stage 1 protocols are revised and transformed into stage 2 Registered Reports, what we would conventionally call a research article, they lose some of this detail. What, then, can we make of such a genre that seemingly serves two functions? Indeed, the genre does serve two functions, one is the traditional business of reporting research findings, and it also responds to the new rhetorical situation of the replication crisis and its demand for a methodological, not results-oriented research article.

Together, this two-stage peer review process changes the scientific research article genre by shifting the rhetorical situation from the reporting of original *results* to the reporting of a research study, inclusive of results. Put simply, the shift occurs from a focus on results to a focus on methodological design and practice (Mehlenbacher 2019b). The two texts produced serve distinct functions during the review process, and thus the genre system of the research article is expanded in a significant manner with the addition of the pre-registered report that transforms into the complete research article in the second stage of review. Reporting results might have been the most obvious exigence for a research article in the traditional model, an exigence that has taken shape over time, as Bazerman (1988) demonstrates. In the Registered Report model, the movement is away from sharing findings (and the inevitable rise of promotional features described by Berkenkotter and Huckin 1995) to the way those results were found in this new two-stage model of peer review. That is, Registered Reports shift us back to how researchers arrived at results they found, rather than focusing on the results themselves. Attending to the process rather than the product helps shift the research article's function to reporting science, which sometimes offers negative or ambiguous results, and those too serve important functions in the development of sound scientific principles.

Maintaining features of the research article genre, Registered Reports continue to serve a number of important genre functions. First among them is often overlooked, that is, the archival function. Here we do not refer to self-archiving to help make one's research more accessible, although we agree that is an important movement, particularly when archived in institutional repositories. Rather, we refer to a longer view of scientific research and development that the research article serves within a scientific field. Researchers some decades out may wish to replicate a study or build on a study, and although a data repository may help in that effort – provided the data are still available, as well as tools for easily accessing that data – it is not the data or even the results that are central to advancing such work. Instead, the methods of a scientific effort are the foundational part of a study. The streamlined, multi-purpose genre functions of the research article mean that, in addition to sharing results, the article contextualizes previous studies and findings and builds on its own theoretical and methodological developments. Thus, a research article serves to document a snapshot of a field's debates, and also to develop a situated argument in the form of an approach and methods. To fully understand the methods, it is important to understand that they are necessarily situated in this longer conversation taking place within a discipline.

4. Conclusions

Conjecture about how and why genres might change with the affordances of the web implicate broader questions about how the web reconfigures knowledge sharing, information-seeking behaviour, and so on. Proponents of registered models will often blog, Tweet, and share video lectures in an effort to encourage others to take up their cause. Journals' review systems are expedited by the affordance of nearly-instant communication online. This is to say that, although in some ways the web did not seem to have much influence on the article directly, beyond the contextualizing and framing of the article, this is rather a product of more complex changes that technologies bring out in our social arrangements. While it may be satisfying to point to particular mechanisms of an interface or tools for sharing results or data, it is instead the complex interaction of socially-oriented technologies that have influenced the research article in important ways. When we talk about how traditional genres have been influenced by the digital age, then, we might rather focus on the temporal rather than the technical orientation in the phrase. The replication crisis demonstrates that the complexity of how old and new genres and technologies interact is necessarily dependent upon the community of genre users. Indeed, this is especially well illustrated by the question of whether or not science *in toto* faces a replication crisis. Qualifying this issue with "psychological and life sciences" is meant to contain the arguments in this chapter to what more clearly seems to be the case of a replication crisis, although certainly the matter is not settled among researchers in these fields. Looking more broadly to physical sciences, the matter of a replication crisis is something more complicated. *Chemistry World* reports that an "unconventional journal" called *Organic Syntheses* has commissioned independent researchers to replicate studies submitted to their journal, and the journal began this practice nearly 100 years ago (Chawla 2017). Although such an approach to review is unconventional, the article goes on to detail why chemistry is a "sweet spot" for replications, since much of the equipment would be available in a number of labs, unlike more specialized fields of the physical sciences requiring enormously expensive equipment. Perhaps in such a case, the fundamental shift in the peer review process would look something different than Registered Reports, and rather similar to the process used by *Organic Syntheses*. What we are underscoring here is that, although the scientific research article has an impressive degree of stabilization, and also typifications across a number of fields and disciplines in the sciences, attending to the particulars of each specialized community of researchers can provide important insights to genre scholars attempting to chart genre change.

Another trajectory emerging from the question of how traditional and new or emerging genres connect to one another through the scientific research article is what such relationships tell us about how communities of practice are being reconfigured. For science, the question of who is sanctioned to speak is one that has been interrogated by science studies scholars for decades. How do we acknowledge public expertise? What is the role of amateur scientists? How can the data collected and analyzed by citizen scientists – everyday people participating in scientific research – be assessed? How does the increasing political engagement by scientists shape the forms of communication that scientists are using? Genre researchers have a particularly salient concept to explore how reconfigurations of experts and publics are unfolding. At the heart of the tension between the specialization of science and the turn toward more public science, the scientific research article becomes an important site of negotiation. Registered Reports provide an example of how new media forms indeed shape the scientific research article in a dramatic reconfiguration, but one mostly internal to science. How new media forms are reconfiguring the scientific article is, then, a rather limited scope. We might instead look to how new media forms dramatically reconfigure social relationships, such as those between a scientific community and broader publics, to better understand what is at stake for those genres emerging and evolving in new media environments. The public discussion of the replication crisis, however, is equally illuminating in that a methodological problem constituted by and embedded within research process genres has become a topic of some public debate (Mehlenbacher forthcoming 2019c).

What we can learn from the case of the scientific research article is about the interplay between traditional genre forms and their online counterparts within a professional sphere of discourse. In these spheres, traditional genres have served important functions in the development of a profession, and in the case of the scientific research article we indeed have an example foundational to the efforts of the profession itself. In such cases, the needs of the discourse community continue to shape the genre in terms of either its stabilization or evolution more so than new affordances offered by the web. Wholesale reinvention of the research article is unlikely as the rhetorical demands that the research article meets continue to shape the profession, in terms of research efforts, and also in terms of the profession as a practice embedded in universities and associated research bodies, such as funding agencies. Augmentation of scientific research articles, such as new features or corresponding platforms for sharing data or the articles themselves, are all important indicators that there is evolution occurring within the profession but perhaps is a less obvious indicator of genre change. Yet, there is genre evolution occurring in the research article, and although it may not be obvious from the product published in its final form as an IMRAD-based research article, change might be found behind the scenes in the process genre users follow to create those products. For genre researchers, then, it is important to continue to attend to activities that constitute the genres we study – the rhetorical situations that call them into being, the processes of inscription that produce typifications, as those typifications may, in fact, serve multiple and changing rhetorical functions.

References

Ball, Philip. 2017. "It's Not Just You: Science Papers Are Getting Harder to Read." *Nature*, 30 March 2017. Last accessed June 15, 2019. https://www.nature.com/news/it-s-not-just-you-science-papers-are-getting-harder-to-read-1.21751

Bazerman, Charles. 1988. *Shaping Written Knowledge: The Genre and Activity of the Experimental Article in Science.* Madison, WI: University of Wisconsin Press.

Berkenkotter, Carol, and Thomas N. Huckin. 1993. "Rethinking genre from a sociological perspective." *Written Communication* 10(4), 475–509.

Berkenkotter, Carol and Thomas N. Huckin. 1995. *Genre Knowledge in Disciplinary Communication: Cognition/Culture/Power.* Hillsdale, NJ: Lawrence Erlbaum.

Casper, Christian F. 2016. "The Online Research Article and the Ecological Basis of New Digital Genres." In *Science and the Internet: Communicating Knowledge in a Digital Age*, ed. by Alan G. Gross, and Jonathan Buehl, 77–98. Amityville, NY: Baywood Publishing.

Chambers, Christopher D. 2013. "Registered Reports: A New Publishing Initiative at Cortex." *Cortex* 49 (3): 609–610. https://doi.org/10.1016/j.cortex.2012.12.016

Chambers, Christopher D., Zoltan Dienes, Robert D. McIntosh, Pia Rotshtein and Klaus Willmes. 2015. "Registered Reports: Realigning Incentives in Scientific Publishing." *Cortex* 66: A1–A2. https://doi.org/10.1016/j.cortex.2015.03.022

Chawla, Dalmeet S. 2017. "Taking on Chemistry's Reproducibility Problem." *Chemistry World*, 20 March 2017. Last accessed June 15, 2019. https://www.chemistryworld.com/news/taking-on-chemistrys-reproducibility-problem/3006991.article

Fahnestock, Jeanne. 1986. "Accommodating Science: The Rhetorical Life of Scientific Facts." *Written Communication* 3 (3): 275–296. https://doi.org/10.1177/0741088386003003001

Garzone, Giuliana. 2012. "Where Do Web Genres Come from? The Case of Blogs." In *Evolving Genres in Web-mediated Communication*, ed. by Sandra Campagna, Giuliana Garzone, Cornelia Ilie, and Elizabeth Rowley-Jolivet, 217–242. Bern, Switzerland: Peter Lang.

Giltrow, Janet, and Dieter Stein (eds.). 2009. *Genres in the Internet: Issues in the Theory of Genre.* Amsterdam: John Benjamins. https://doi.org/10.1075/pbns.188

Gross, Alan G., and Joseph E. Harmon. 2016. *The Internet Revolution in the Sciences and Humanities.* Oxford, UK: Oxford University Press.

Gross, Alan G., Joseph E. Harmon, and Michael Reidy. 2002. *Communicating Science: The Scientific Article from the 17th Century to the Present.* Oxford, UK: Oxford University Press.

Harmon, Joseph E. 2016. "The Scientific Journal: Making It New?" In *Science and the Internet: Communicating Knowledge in a Digital Age*, ed. by Alan G. Gross, and Jonathan Buehl, 33–58. Amityville, NY: Baywood Publishing.

Herring, Susan C., Lois A. Scheidt, Sabrina Bonus, and Elijah Wright. 2005. "Weblogs as Bridging Genre." *Information, Technology & People* 18 (2): 142–171. https://doi.org/10.1108/09593840510601513

Kaplan, Sarah, and Joanna Radin. 2011. "Bounding an Emerging Technology: Para-Scientific Media and the Drexler-Smalley Debate about Nanotechnology." *Social Studies of Science* 41 (4): 457–485. https://doi.org/10.1177/0306312711402722

Kelly [now Mehlenbacher], Ashley R., and Meagan Kittle Autry. 2013. "Access, Accommodation, and Science: Knowledge in an 'Open' World." *First Monday* 18 (6): 10pp. Last accessed June 15, 2019. https://firstmonday.org/ojs/index.php/fm/article/view/4341/3684doi:10.5210/fm.v18i6.4341

Kittle Autry, Meagan. 2013. Genre Change Online: Open Access and the Scientific Research Article Genre. Raleigh, NC: North Carolina State University. Unpublished doctoral dissertation.

Li, Li-Juan, and Guang-Chun Ge. 2009. "Genre Analysis: Structural and Linguistic Evolution of the English-Medium Medical Research Article (1985–2004)." *English for Specific Purposes* 28 (2): 93–104. https://doi.org/10.1016/j.esp.2008.12.004

Mackenzie Owen, John. 2007. *The Scientific Article in the Age of Digitization.* Dordrecht, Netherlands: Springer.

McNeill, Laurie. 2003. "Teaching an Old Genre New Tricks: The Diary on the Internet." *Biography* 26: 24–47. https://doi.org/10.1353/bio.2003.0028

Mehlenbacher, Ashley R. 2019a. *Science Communication Online: Engaging Experts and Publics on the Internet.* Columbus, OH: The Ohio State University Press.

Mehlenbacher, Ashley R. 2019b. "Registered Reports: An Emerging Scientific Research Article Genre." *Written Communication* 36 (1): 38–67.

Mehlenbacher, Ashley R. forthcoming 2019c. "Exploring Conversations about Science in New Media." In *Routledge Handbook of Language and Science*, ed. by Lynda Walsh and David Gruber (in press). New York, NY: Routledge.

Mehlenbacher, Ashley R., and Carolyn R. Miller. 2017. "Intersections: Scientific and Parascientific Communication on the Internet." In *Landmark Essays on the Rhetoric of Science: Case Studies*, 2nd edition, ed. by Randy A. Harris, 239–260. New York, NY: Routledge.

Miller, Carolyn R. 1984. "Genre as Social Action." *Quarterly Journal of Speech* 70, 151–167.

Miller, Carolyn R., and Jeanne Fahnestock. 2013. "Genres in Scientific and Technical Rhetoric." *Poroi* 9 (1): 12. https://doi.org/10.13008/2151-2957.1161

Miller, Carolyn R., and Ashley R. Kelly [now Mehlenbacher] (eds.). 2017. *Emerging Genres in New Media Environments*. London, UK: Palgrave Macmillan. https://doi.org/10.1007/978-3-319-40295-6

Miller, Carolyn R., and Dawn Shepherd. 2004. "Blogging as Social Action: A Genre Analysis of the Weblog." In *Into the Blogosphere: Rhetoric, Community, and the Culture of Weblogs*, ed. by Laura Gurak, Smiljana Antonijevic, Laurie Johnson, Clancy Ratliff, and Jessica Reyman, 1–21. Minneapolis, MN: University of Minnesota Libraries. Last accessed June 15, 2019. http://hdl.handle.net/11299/172818

Miller, Carolyn R., and Dawn Shepherd. 2009. "Questions for Genre Theory from the Blogosphere." In *Genres in the Internet: Issues in the Theory of Genre*, ed. by Janet Giltrow, and Dieter Stein, 263–290. Amsterdam: John Benjamins. https://doi.org/10.1075/pbns.188.11mil

Nosek, Brian, et al. 2015. "Estimating the Reproducibility of Psychological Science." *Science* 349 (6251): aac4715.

Nwogu, Kevin N. 1997. "The Medical Research Paper: Structure and Functions." *English for Specific Purposes* 16 (2): 119–138. https://doi.org/10.1016/S0889-4906(97)85388-4

Schmidt, Charles W. 2009. "Communication Gap: The Disconnect between what Scientists Say and what the Public Hears." *Environmental Health Perspectives* 117 (12): A548–A551. https://doi.org/10.1289/ehp.117-a548

Schryer, Catherine F. 1993. "Records as Genre." *Written Communication* 10: 200–234. https://doi.org/10.1177/0741088393010002003

Schulson, Michael. 2018. "Science's "Reproducibility Crisis" Is Being Used as Political Ammunition." *Wired*, 20 April 2018. Last accessed June 15, 2019. https://www.wired.com/story/sciences-reproducibility-crisis-is-being-used-as-political-ammunition/

Somers, James. 2018. "The Scientific Paper Is Obsolete." *The Atlantic*, 5 April 2018. Last accessed June 15, 2019. https://www.theatlantic.com/science/archive/2018/04/the-scientific-paper-is-obsolete/556676/

Swales, John M. 1990. *Genre Analysis: English in Academic and Research Settings*. Cambridge: Cambridge University Press.

Swales, John M. 2004. *Research Genres: Explorations and Applications*. Cambridge, UK: Cambridge University Press. https://doi.org/10.1017/CBO9781139524827

Yong, Ed. 2010. "On Jargon, and why It Matters in Science Writing." *National Geographic*, 24 November 2010. Last accessed June 15, 2019. https://www.nationalgeographic.com/science/phenomena/2010/11/24/on-jargon-and-why-it-matters-in-science-writing/

CHAPTER 4

The graphical abstract as a new genre in the promotion of science

Graciela Rabuske Hendges and Cristiane Salete Florek
Universidade Federal de Santa Maria

This study explores an increasingly popular Internet genre, the graphical abstract, and its relations to verbal abstracts and to the images used in research articles. We develop a preliminary description of the form and function of GAs, based on a sample of texts from chemistry and engineering and on journal editors and authors' perceptions about the genre. The results show significant variation among the texts, yet some discipline-internal regularities could be observed. These findings do not necessarily reflect editors and authors' perceptions. The frequent use of color and drawings is related to the promotional role of graphical abstracts in attracting a wide range of readers, which approximates the genre to verbal abstracts, but dissociates it from scientific images. Graphical abstracts seem to combine the best of its precedent part-genres to optimize the journal readership.

Keywords: graphical abstract, Internet genres, genre relations, verbal abstract, images in science, commodification of scientific discourse

1. Introduction

As also stated in previous chapters, genres have been defined as relatively stable uses of language in recurrent social activities that develop within specific sociocultural contexts (Swales 1990; Motta-Roth and Heberle 2015). Changes in existing genres and the emergence of new genres are directly associated to changes in social activities and new communicative needs, and are frequently triggered by technological developments.

The dynamic nature of genres has raised a number of questions in the field of genre theory, particularly after the emergence of digital genres. Some of these questions are related to acceptable genre variation;[1] "unacceptable violation" of a

1. Genre variation can refer both to historical evolutionary changes in a genre as well as to contextual variation in a genre (differences found in research articles across different disciplines).

https://doi.org/10.1075/pbns.308.04hen
© 2019 John Benjamins Publishing Company

genre; criteria for sanctioning or rejecting genre variation; authority, motivations and means for creating new genres; criteria for classifying genres as hybrids in the process of generic change (could all genres be studied as hybrids of previously existing antecedent genres?) (Devitt 2004: 218). These questions are relevant even in the context of science, where apparently stabilized genres such as the research article (Pérez-Llantada 2013) have been increasingly adapted to the affordances of e-journals, for example, presenting audiovisual data and hyperlinks that rearrange the canonical linear IMRD (Introduction-Methods-Results-Discussion) pattern into nonlinear blocks, encouraging different routes to access scientific information (see Harmon this volume). These changes respond to both scientists' interests in providing more visually detailed data to support their argument – particularly in disciplines where scientific images play a central role – but most importantly, to journal publishers' interest in satisfying multiple readers and reading demands.

On a more critical note, the changes in scientific communication can be related to the broader pressures of marketization – "the extension of market models to new spheres" (Fairclough 1992: 99) – and commodification – the organization and conceptualization of social domains, "whose concern is not producing commodities in the narrower economic sense of goods for sale," in terms of commodity production, distribution and consumption (Fairclough 1992: 207). Scientific activity has been increasingly affected by these forces (Fairclough 1992, 1995; Pérez-Llantada 2012) and competition for research space and for readership "has never been fiercer" as "participation in the global exchange of information is now a prerequisite for promotion and job security for a growing number of academics around the world" (Hyland 2009: 18). While editors and authors compete to increase the visibility of their publications, readers have been challenged to develop selective skills to deal with vast amounts of potentially interesting literature.

Some fields may be less affected than others by these trends (Swales 2004), but there seems to be a consensus about the role of digital technologies in intensifying the commodification process by means of, for example, the widespread proliferation of electronic journals (e-journals) and online academic libraries (Gu and Blackmore 2016), whose maintenance and survival depends on entering the battle for *consumers* rather than readers (consumerism as "a shift, or apparent shift, in power from producers to consumers" (Fairclough 1992: 109) is another trend within marketization and commodification). According to Fairclough (1995: 139), in discursive terms, this shift causes "the generalization of promotion as a communicative function" across discursive practices and a consequent "restructuring of boundaries" between advertising and these practices, often leading to the creation of "new hybrid partly promotional genres" (Fairclough 1995: 139).

In the context of academic discourse, a number of new freely-available genres (or part-genres) have been introduced such as the *author summary*, the *lay summary*, the *highlights* and the *significance statements*. In these texts, promotionalism is construed in their aim of broadening readership of e-journals to non-expert audiences (Breeze 2016). But promotionalism has also been perceived in traditional genres which are *par excellence* considered non-promotional, such as the scientific journal article (Ayers 2008) and the journal article abstract (Huckin 2001; Ayers 2008). In *The Article of the Future* editorial project (see Harmon this volume; Luzón and Pérez-Llantada this volume), launched by Elsevier in 2009, the affordances of new technologies for the presentation and distribution of research findings (wider readership, incorporation of hyperlinks, multimedia and supplementary files in online research articles) have been advocated as promotional mechanisms that make articles more user-friendly, improve navigation and enrich content (Elsevier 2012). Not rarely a number of these changes give particular prominence to images as a major tool to attract readers. Fairclough (1995: 139) sees shifts "in the relative salience of different semiotic modalities" as the colonizing effect of advertising – with its "greater dependence upon visual images at the relative expense of verbal semiosis" – over other discourses.

The graphical abstract (henceforth GA) is an example of a new online scientific genre[2] that has been defined by the use of eye-catching visuals to attract reader attention and increase journal visibility (*Nature Chemistry* 2011). The rapid spread of GAs across a number of fields suggests that it has been successful in accomplishing these goals, but, so far, there is no research evidence on what linguistic and/or visual features are the most effective in this process. Equally uncertain is the extent to which the function and form of GAs differ from that of traditional verbal abstracts and of the images used in the research articles.

In this scenario, the present work attempts to build knowledge on the issues of genre synergies and relations, and motivations for genre stability and change, through a preliminary analysis of the form and function of the GA in two fields where it has been used steadily: chemistry and engineering. Given the lack of detailed examinations of GAs in previous literature and the novel character of the genre, we consider a small sample of texts to offer a general picture of GAs as well as details that could help in formulating more precise questions in future research. We complement our textual analysis with a contextual analysis based on e-journal publishing policies and on perceptions and practices of journal editors and of authors/readers about the form and function of the GA.

2. For the purpose of the present work, we will refer to the graphical abstract as a genre because it is also published independently, separated from the research article it originally accompanies.

Following Todorov's principle (1976: 15) that "[a] new genre is always the transformation of one or several old genres," our discussion also explores how the GA relates to the form and function of the traditional abstract and of the images used in research articles to determine what rhetorical patterns from research article abstracts and visuals are inherited by GAs and what is new. We assume that the GA is not a mere reproduction of existing genre elements but a digital genre with new form and purpose. As pointed out by Miller and Shepherd (2004: 2), "when a type of discourse or communicative action acquires a common name within a given context or community, that's a good sign that it's functioning as a genre."

In light of these considerations, our research questions are:

1. How are GAs shaped in terms of form?
2. What is the communicative purpose of the GA?
3. How does the GA relate to the form and function of its antecedents – the traditional verbal abstract and article images?
4. What linguistic/visual features in GAs can be linked to a promotional goal?
5. What are journal editors and authors/readers' perceptions about the GA?

We will begin by reviewing theoretical knowledge about two GA precedents: traditional verbal abstracts and research article visuals, followed by a brief description of the GA.

2. Existing research article elements: Abstracts and visuals

Journal article abstracts and visuals have frequently been highlighted as playing central roles in reader persuasion (Swales 1990; Miller 1998; Nieman 2000; Ayers 2008; Swales and Feak 2009; Hemais 2014). For most readers, abstracts work as the front page headlines of newspapers in foregrounding newsworthy information from the research article they accompany (Swales 1990; Ayers 2008). Similarly, expert readers in many fields use the images in an article as a "fast track to the new" (Miller 1998: 43) and frequently skim images before other parts of the research article. Both abstracts and scientific images have highly promotional value and will determine whether an article is worth reading or not.

2.1 Abstracts

The importance of journal article abstracts in scientific communication can be measured by the large number of studies about the genre in a wide range of disciplines. In general, abstracts are defined as "advance indicators of the content and structure of the following text" (Swales 1990: 179) that fulfill two primary intertwined purposes: to provide a synopsis of the accompanying article in a way to promote it. In terms of form, abstracts have been described as reflecting the IMRD pattern of research articles (Swales 1990: 179), but recent research (Huckin 2001; Ayers 2008) has revealed that abstracts have increasingly given salience to results and discussion, while methods are "being relegated to a secondary position" (Ayers 2008: 23). These modifications echo the discourse of popular science news, in which "findings are highlighted, but the means by which the findings were obtained are placed in the background … The reader is not positioned as knowledgeable but as needing to be enticed into the article" (Miller 1998: 31). Manipulations of the IMRD structure along with the use of the present tense in the active form and the insertion of explicit definitions in abstracts (Ayers 2008: 23) have been interpreted as "a kind of 'democratization' of the scientific community, with an opening up to the general public" (Ayers 2008: 39). This tendency involves reducing "overt markers of power asymmetry between people of unequal institutional power" (Fairclough 1992: 98).

2.2 Journal article visuals

The function and form of images in research articles has been considered more frequently in fields outside academic discourse theory. Johns (1998: 183) sees at least two reasons for this: applied linguists "are most interested in, and comfortable with, written language" and "are trained in the humanities, where words are central to disciplinary values and argumentation." In spite of that, studies of academic discourse consistently acknowledge the importance of scientific visual meanings as key elements in building and supporting the scientific argument and in advancing science (Johns 1998; Miller 1998; Swales 2004; Hemais 2014). Even before e-journals, the importance and size of scientific visuals grew steadily, in spite of cost and space limitations of the printed page (Miller 1998). That is the case in essentially visual subjects such as chemistry, where images in the form of chemical structures are irreplaceable (Nature Chemistry 2011).

The power of images in providing evidence to sustain scientific argument can be encapsulated by their unique capability for compressing complex concepts, relations, and an enormous amount of hard data in compact spaces, so that meanings can be captured *at a glance*. In doing so, images can give the illusion of direct access to the data as if nature was speaking for itself and "as if the data rather than the scientist are carrying the argument" (Miller 1998: 30–31). These properties of images can be enhanced with the possibilities brought by online journals (see Harmon this volume), as described in *The Article of the Future* project: "imagine moving from a traditional flat map view of a terrain to an interactive satellite view," "imagine switching from viewing proteins as a static image or a 4-character code to a 3-dimensional view" and "imagine changing from [...] manual measurement of values within a graph to being able to analyze them with absolute accuracy and achieve more qualified results using an interactive crosshair viewer" (Elsevier 2012). *The Article of the Future* project gives special attention to images as major elements of research articles to be enhanced with technological resources. It is interesting to note that these resources are emphasized as tools for enriching reading experiences rather than writing and this is another evidence of consumerism[3] as a general tendency associated to the use of technology in academic communication.

The visual persuasion of scientific images is intimately connected to disciplinary values, tools and symbolism. In most fields, the production and presentation of scientific images depends on highly specialized technology (satellites, for example) and expert visual literacy to manipulate hardware and software tools and symbolic language. Similarly, the interpretation of these images requires a significant degree of technical expertise. For the academic discourse analyst, the lack of technical literacy challenges the study of scientific images and encourages the inclusion of expert informants in the interpretation of visual meanings.

The relevance of images in research publication has become more evident with the birth of GAs. The following sections explore the evolution of the GA and the extent and means by which this digital genre inherits features from verbal abstracts and from research article images.

3. The graphical abstract as a digital genre

When we started this project, we believed that the GA was a new digital genre without a print counterpart. To our surprise, we soon found out that the origin of GAs dates back over 40 years, specifically to 1976, when it was first published in the then multilingual chemistry journal *Angewandte Chemie* (Nature Chemistry 2011; Lane, Karatsolis, and Bui 2015) (Figure 1).

Thus, "far from being driven by the rise of the internet and the opportunities it affords scholarly publishing" (Nature Chemistry 2011: 571), the initial motivations for the appearance of GAs are linked to the visual essence of the field of chemistry and to the role of symbolic chemical language as an international code for readers of various linguistic backgrounds (Lane, Karatsolis, and Bui 2015). Nevertheless, given the inexistent previous literature about the early *Angewandte Chemie* GAs and their regular use only in this journal, it seems difficult to make generalizations about the function and form of the genre in this early stage, before e-journals.

3. Fairclough (1992: 117) points out that it may be arguable to what extent the shift in power from writer/producer to reader/consumer "is substantive or cosmetic."

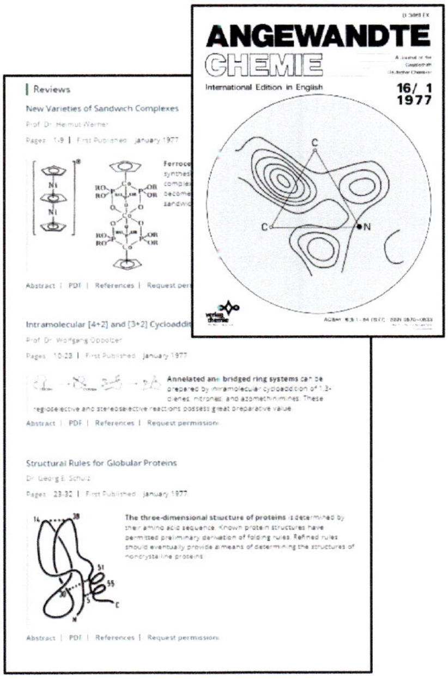

Angew.Chem. 16:1, 1977

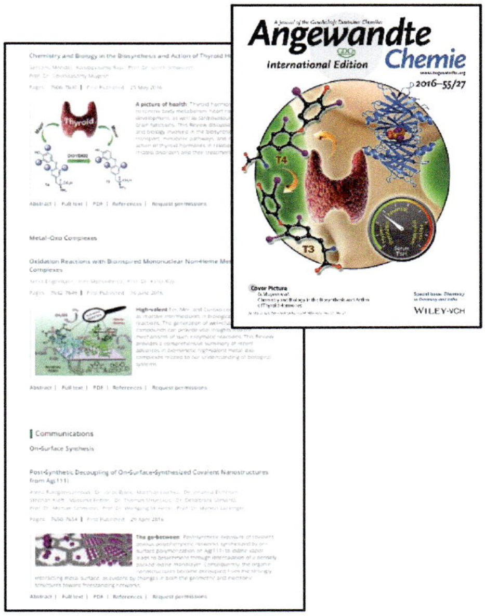

Angew.Chem. 55:27, 2016
(graphics in colour in the original)

Figure 1. GAs in a 1977 issue of *Angewandte Chemie* (left) and in a 2016 issue (right)

Since the mid-2000s, the GA has spread to an increasing number of fields, "either as a requirement or suggestion" of journals (Hullman and Bach 2018: 1). In the social sciences, for example, Yoon and Chung (2017) mapped a 350% rise in GA use from 2011 to 2015. This popularity has inspired studies about the genre (Lane, Karatsolis, and Bui 2015; Yoon and Chung 2017; Hullman and Bach 2018) and a flourishing number of less formal discussions about the purpose and effectiveness of GAs (Nature Chemistry 2011; Lowe 2012; Cox 2015; Gilaberte et al. 2016) as well as of GA design guidelines. Nonetheless, most of the information available in these documents is limited to decontextualized generalized statements about the function of GAs, as pointed out by Lane, Karatsolis, and Bui (2015), and to technical details about file type (TIFF, PDF, etc.), quality and resolution (pixels, dpi) of images and physical dimensions on the page (height, width). Authors' and readers' perceptions, behaviors and needs when creating and reading GAs are yet unknown.

Overall, the contemporary GA is defined as an online genre in the form of a visual summary of the content of the article it accompanies (Lane, Karatsolis, and Bui 2015; Yoon and Chung 2017; Hullman and Bach 2018). It has been characterized as having the two-fold primary purpose of visually and persuasively illustrating the journal article's main idea (Lane, Karatsolis, and Bui 2015). In other words, the purpose of GAs is "to distil the take-home message of an article" into an image that grabs the reader's attention and makes him "want to read the paper it is advertising" (*Nature Chemistry* 2011). Another less central role of GAs is to make authors reflect about their work and understand its fundamental conclusion (Cox 2015; Gilaberte et al. 2016).

These features of GAs show a close relationship with the purpose of traditional abstracts (Swales 1990), but also that the main difference between them is that GAs use the visual semiotic mode to convey meaning, which, by nature, uses a totally different lexicon (shapes, volumes) and grammar (connecting devices, position, angles, axis, framing, size, etc.) based on a logic of space rather than time to organize content. This affords the packing of information in such a way that "a whole picture can be 'taken in' in an instant" (Painter, Martin, and Unsworth 2012: 91).

In combining the summarize-to-attract purpose of traditional abstracts with the visual packing potential of images that can be captured "at a single glance," the GA can speed up the promotional role of verbal abstracts. In the present work we explore these genre relations.

4. Methodology

The data for the present study were collected from a textual sample of 30 GAs from the fields of chemistry and engineering. Both fields have the longest tradition in publishing GAs and may offer a more stabilized description of the genre. The texts were selected from 15 high impact factor journals where the GA was a mandatory feature of manuscript submission. Within the list of journals, we found that five titles were interdisciplinary, being relevant for both chemistry and engineering. The resulting list includes the following: (1) chemistry: *Chemical Reviews*; *Angewandte Chemie*; *Chemical Science*; *The Journal of Physical Chemistry Letters*; and *Organic Letters*; (2) engineering: *Acta Biomaterialia*; *Acta Materialia*; *Insect Biochemistry and Molecular Biology*; *Scripta Materialia*; and *Powder Technology*; (3) interdisciplinary: *Energy and Environmental Science*; *Nano Letters*; *ChemSusChem*; *Applied Catalysis B, Environmental*; and *Journal of Catalysis*.

For corpus selection, we considered GAs from the most recent issue in each journal by June 30, 2016, exclusively from sections that publish articles, resulting in three sets of GAs: 10 from chemistry, 10 from engineering and 10 from interdisciplinary journals. Some journals publish only review articles (e.g. *Chemical Reviews*), others publish only experimental research articles (e.g. *Scripta Materialia*) and a third group publishes both types (e.g. *Angewandte Chemie)*. When possible, we collected one GA from each section, the first to appear in the table of contents. However, when the journals in our study published only one type of article, we selected the first and last GA appearing in the table of contents. The resulting sample included 23 GAs from experimental research articles and 7 GAs from review articles, two types of research articles that differ significantly in form and function (Swales 2004). In spite of this unbalance in the corpus, we decided to maintain the sample for comparative purposes and because it is out of the scope of the present study to offer a systematic genre analysis of GAs. Therefore, we do not expect to offer a generalizable description of GAs nor do we expect this would be possible even with a larger sample, given that the GA does not seem to be stabilized enough at the current stage of development.

Our analytical procedure follows a text-first approach complemented with contextual information about practices and perceptions of GA publication, production and interpretation (Askehave and Swales 2001: 207). Five interconnected and recursive steps were implemented:

1. description of GAs in terms of four categories: *layout* and *visual entities* (the number of visual entities that constitute the GA and how they are positioned in the space of the GA), *originality* (whether the GA is an original visual or partly uses images from within the article), and *nature of the images* (whether images are more realistic as in photographs or more abstract as in chemical schemes). The analytical categories and coding criteria were largely based on previous work (Lane, Karatsolis, and Bui 2015; Yoon and Chung 2017; Hullman and Bach 2018), but also adapted when necessary;
2. interpretation of observed patterns to identify the communicative purpose/function of GAs;
3. assessment of the context of production and of reception of GAs, through editorial policy documents and through the participation of specialists from each field – editors and authors/readers;
4. reinterpretation of the purpose of GAs based on contextual data;
5. review of the status of the GA as a digital genre and its relation to verbal abstracts and research article visuals.

The participants were included in the study via online questionnaires. Journal editors were asked 12 questions (eight multiple choice and four open-ended) about professional experience, motivations for adopting GAs in the journal, the purpose of GAs, perceptions about what makes a good GA, the impact of GAs on the journal and future perspectives. Authors/readers were asked 29 questions (22 multiple choice and seven open-ended). Questions 1–7 addressed the participants academic profile; questions 8–16 explored previous experiences in GA writing and reading; questions 17–29 explored perceptions about the form and function of GAs. We obtained a response rate of 11.8% of the authors/readers (11 of 93 contacted (co)authors completed the questionnaire) and 6.8% of the editors (5 of

73 editors returned). The apparently low response rates did not prevent us from considering this data set as it provided helpful information for the interpretation of the textual analysis, considering that proposing generalizations is out the scope of our analysis.

5. Results and discussion

The analytical categories used for describing textual patterns in our sample of GAs – *layout and visual entities*, *originality*, and the *nature of the images* – were initially chosen based on the three available analyses of GAs (Lane, Karatsolis, and Bui 2015; Yoon and Chung 2017; Hullman and Bach 2018). They seemed to be useful tools to identify evidence about the purpose of GAs and to evaluate the status of GAs in relation to potential synergies with verbal abstracts and with the images in the body of the research article.

5.1 Layout and number of visual entities in GAs

In the context of GAs, layout design is defined as "the organization of graphical elements in the 2D space" and is described based on the amount of visual "entities or nodes" that a GA contains (Hullman and Bach 2018: 5–6). Because the authors provide no definition for *visual entities*, we related their use of the term to the concept of "focus group" proposed by Painter, Martin, and Unsworth (2012: 91–92): "material that is grouped together compositionally as some kind of unity or 'eyeful' to which we are guided to attend."

The contribution of layout analysis in identifying the purpose of GAs and mapping genre relations is that it can reveal whether the form of GAs comprises one or multiple visual entities. This has a direct impact on the purpose that has been attributed to GAs by the literature and author guidelines: summarizing the research article in an image that can be captured at a glance. We assume that the more visual entities a GA has, the more difficult it is to read it rapidly at once, in less time than a verbal abstract would demand.

To develop our layout analysis, we mapped the number and spatial distribution of focus groups in each GA and explicit indicators of reading paths based on the nine types of layout found by Hullman and Bach (2018). The authors examined the presence of visual entities and of explicit connecting devices in a sample of fifty-four GAs from a variety of disciplinary fields. According to these authors, GAs use layouts that can be classified along a continuum ranging from *linear* layouts – "that convey an explicit reading order" with use of explicit visual cues (namely *linear*, *circular*, *zig-zag*, *forking* and *nesting*) – to *free* layouts – "that can be read in various orders" (namely *parallel, orthogonal, centric* and *free* layouts) (Hullman and Bach 2018: 6). We found that all the GAs in our sample fit into at least one of the eight layout patterns described by Hullman and Bach (2018), thus it was not necessary to propose new categories (Table 1).

From the point of view of number of visual entities, 22 GAs (73.3%) show from two to eight elements. In 14 of those (6 *linear*; 4 *forking*; 3 *nesting*, and 1 *circular*), there is explicit signaling of reading paths using directionality arrows.[4] When arrows are absent, as in *parallel* layouts, the horizontal or vertical side-by-side placement of visual entities suggests reading paths that may either go from left to right or from top to bottom following traditional reading orders in western culture. Alternatively, reading paths may follow readers' interests, usually shaped by disciplinary values and conventions (for example, the expert reader in chemistry may be more attracted by a chemical scheme than by a photograph).

We do not have enough evidence to draw conclusions about the relation between differences in layout patterns and type of GAs (review or experimental), nor about preferred layouts and disciplinary practices. It is possible to argue, however, that the predominant use of multiple visual entities and explicit signaling of reading trajectories is related to the logic of verbal language in which blocks of information are accessed one at a time. This may delay the reading process and interfere with the purpose of the GA to allow quick identification of relevant articles.

The examination of author guidelines in the 15 journals revealed no explicit layout specifications, but two journals offer recommendations that can be linked to the number of visual entities to be used. *Acta Materialia* states that the GA "visualizes *one* process" and that authors should "try *to reduce distracting and cluttering elements* as much as possible." *Organic Letters* recommends the GA "should *be simple* [...] and should be in the form of structure, graph, drawing, photograph, or scheme – *or a combination*."

When asked about aspects related to layout, the specialist participants in our study show a range of perceptions. While seven (43.7%) participants indicate the GA should be a single image (= a single visual entity, as in the *free* layout pattern), nine (56.2%) understand it should be a combination of images (= multiple visual entities). On the one hand, these data could indicate that choices of layout depend on the discipline and on the purpose of each individual study. On the other hand, divergences could be interpreted as a result of insufficient information about the most effective layout for each case. Previous studies agree that GA authors are not aware of these possibilities and "arrive at patterns (e.g. choice of layout) without necessarily understanding how a given pattern relates to their communication goals" (Hullman and Bach 2018: 2).

4. When compared to arrows used to connect a visual element to a (verbal) label.

Table 1. Types of layout designs found in the GAs with examples

Cont.	Layout pattern	Definition (Hullman and Bach 2018)	%	Example
LINEAR	Linear	Uses arrows or other visual cues to designate a clear reading order	20	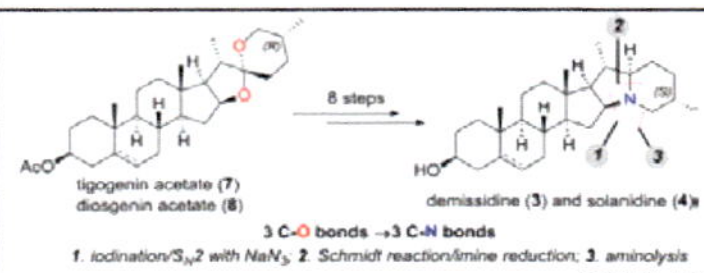 CGA#10
	Circular	Uses arrows or other visual cues to organize information in circular path	3.3	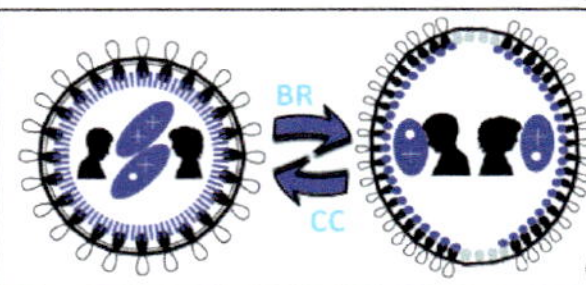 CGA#5
	Forking	Includes both cues to a linear reading order and cues that break the implied linearity	13.3	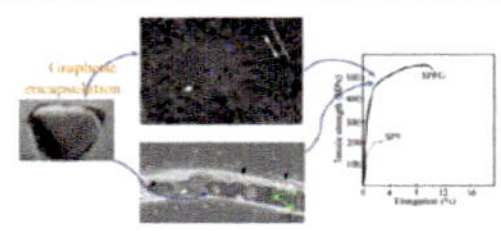 EGA#18
	Nesting	Is analogous to footnoting in text, where a nested frame within the larger GA contributes additional information	10	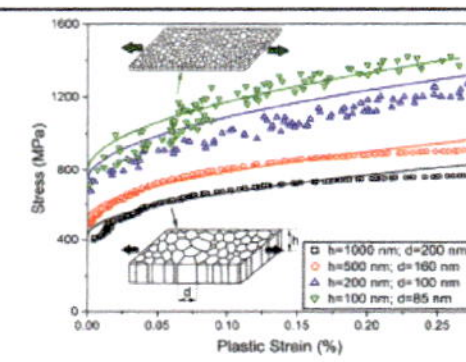 EGA#14
	Parallel	Consists of horizontal or vertical juxtaposed multiple alternatives for comparison, or multiple visual elements juxtaposed without a clear visual cue	20	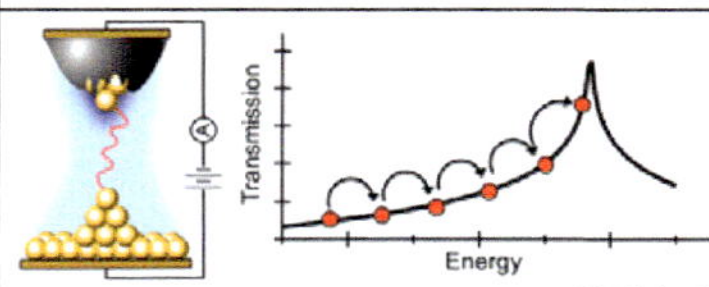 CEGA#24
	Orthogonal	Does not always indicate one clear reading order, but maps different information to both the horizontal and vertical dimensions	3.3	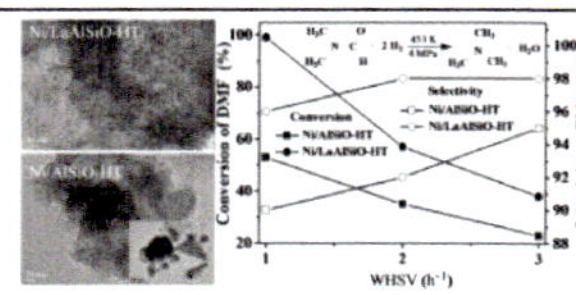 CEGA#29
	Centric	Divides the space into center and periphery, typically mapping elements to both types of position such that it is unclear which direction should be examined after the center	3.3	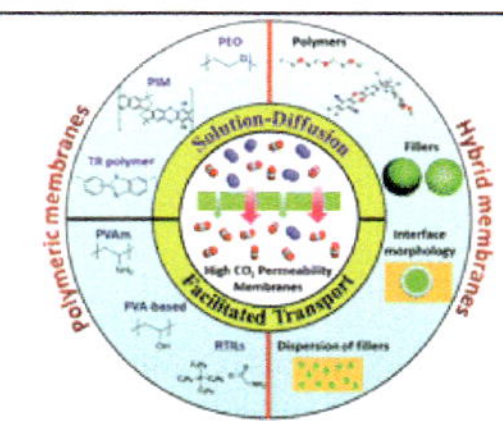 CEGA#21
FREE	Single or Free	Consists of an image that is not clearly differentiable into sub-images	26.6	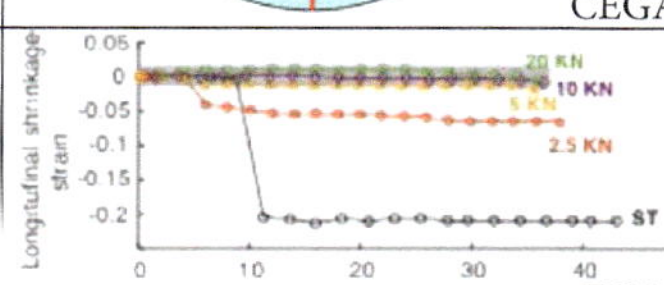 EGA#17

In the next section, we examine if the preferred use of multiple visual entities in GAs is related to reproduction of a combination of images from within the research article.

5.2 Originality

In this section we report results about the originality of the images presented in GAs. Following Yoon and Chung (2017: 1373), we analyzed whether the GAs are *duplicates* of one or more existing images from within the article, *modified* images – when the GA includes an image copied from within the article but subtly modified or integrated with new images, or *new*, when the GA is a "newly created" image.

We see originality as an interesting aspect to analyze from the perspective of genre relations (relations between the GAs and the articles on which they draw, in this case). It shows whether the GA is a reproduction of a visual element from the research article or an adaptation that incorporates new interaction, navigation and multimedia features afforded by the web. Considering that the development of new genres is influenced by antecedent genres, it is important to consider that "those antecedents may sometimes be inappropriate and may limit the new genre's ability to fulfill its functions" (Devitt 2014: 134).

We found that *modified* images are slightly predominant (13 GAs, 43.3%), followed by GAs that *duplicate* an existing image (9 GAs, 30%), and by GAs with *new* images (8 GAs, 26.6%) (Table 2). This result is inversely proportional to the pattern found by Yoon and Chung (2017) in 772 social sciences GAs, as *modified* images were the least frequent in their texts (only 5%), *duplicate* images were predominant (58%) and only 7.38% showed *new* images. Nonetheless, when our results are considered by discipline, engineering GAs show the same pattern as social sciences GAs, with predominance of *duplicate* images, in six GAs (60%), three GAs (30%) with *modified* images and only one GA (10%) that shows a *new* image (Table 2). Compared to engineering, the patterns in chemistry show the exact opposite, as *new* images predominate (5 GAs, 50%), followed by *modified* (4 GAs, 40%) and *new* (1 GA, 10%). In between is the interdisciplinary group, with most GAs with *modified* images (6 GAs, 60%), two (20%) with *duplicate* images and two (20%) with *new* images.

This divergence may be clearly related to the longer history and greater popularity of GAs in chemistry and, consequently, to a clearer understanding by editors and authors in the field of the different purposes of this new genre in terms of information, persuasion and readership when compared to the purpose of images inside an article. This raises the assumption that maybe in this disciplinary context the genre is somehow stabilizing. Publishing policies in chemistry journals overtly encourage authors to "avoid choosing a graphic that already appears within the

Table 2. Originality of the images used in GAs

	Levels of originality					
	Duplicate		Modified		New	
	n	%	*n*	%	*n*	%
Chemistry	1	10	4	40	5	50
Engineering	6	60	3	30	1	10
Interdisciplinary	2	20	6	60	2	20
Totals	9	30	13	43.3	8	26.6

text of the manuscript (*Organic Letters*)." Expectedly, the chemistry specialist participants in the study tend to agree (75%) that the GA should not reuse images from the article, even important images.

It is plausible to assume that this consensus derives from the earlier adoption of GAs in chemistry since 1976, while in the social sciences they have been used since 2010 (Yoon and Chung 2017). Because generic stability is achieved to some extent, it could be assumed that the experimental phase of GAs in chemistry is closer to an end than in other fields. Lane, Karatsolis and Bui (2015) discuss this issue in their analysis of the evolution of GAs in the *Chemical Engineering Journal*, in which initially the GAs used "images pulled from within the articles" ranging from single to multiple images, from 3D diagrams to mathematical modeling, from gray scale to multi-color, and later issues published GAs that consisted of new images. This variation is interpreted as reflecting the initial stage in the development and establishment of the genre, when rapid changes in form and function are expected while users adapt the prototypes to contextual needs until some stability is achieved.

The evolution of a genre "is often but not always a gradual process of subtle adaptations" and "the factors that encourage or inhibit the generic changes probably differ in every circumstance, but they appear to involve always both contextual and individual factors" (Devitt 2004: 134). Therefore, the predominance (6 GAs, 60%) of duplicate images in the corpus of engineering GAs may be linked to an unclear picture about the purpose of the genre and what kind of image accomplishes that purpose. Consequently, authors may be unwilling to do the extra effort or may understand the duplicate image does the job. As in chemistry, it is possible to project that, as the genre evolves, the degree of originality of the images will increase. In this field, the use of different images in research articles and in GAs suggests a difference in purpose: in the first, to provide evidence to convince the reader about the validity of the argument; in the second, to design a visual summary of the study able to convince the reader to read and download the article.

5.3 Nature of the images

The nature of the images used in GAs is another feature that called attention in previous examinations of the genre (Yoon and Chung 2017; Hullman and Bach 2018). It refers to degrees of abstraction of the images used in GAs, ranging from images that "move from the empirical 'real'[5] toward theoretical abstraction" (Bezemer and Kress 2008: 180). Realistic images show correspondence between the depicted reality and "human optical perception" (Lane, Karatsolis, and Bui 2015), such as photographs, while abstract images represent the world in interpreted coded ways, reducing it to essential generalized features, such as chemical schemes (Kress and van Leeuwen 2006). Abstract images are conventionally associated with scientific discourse, while photographs are the standard in everyday communication.

In the present analysis, the nature of the images used in GAs brings important insights about synergies of purpose and form between the GA and its predecessors: the traditional verbal abstracts and the images used in research articles. It shows to what extent the GA inherited the purpose of the verbal abstract to appeal to wider audiences outside the immediate field (Ayers 2008), by suppressing more technical content (e.g. abstract scientific images) and emphasizing content that is accessible to non-experts in the field (e.g. drawings and photographs). Where abstract scientific images predominate in the GAs, the genre relation will be more explicitly established with the images in the research article, which target the expert reader. The previous section indicated that this is the case in engineering GAs, which mostly duplicate images from the research article. In this section we therefore examine whether the GAs "will make scientific publications more accessible and understandable for in- and out-of-domain researchers as well as 'lay' audiences like students, journalists, or members of the public" (Hullman and Bach 2018: 1).

Making science appealing to non-experts through images has been the job of science popularization (Miller 1998; Nieman 2000; Dimopoulos, Koulaidis, and Sklaveniti 2003; Bezemer and Kress 2008). The recontextualization of scientific images in the media and in textbooks has shown a prominence of realistic images that "portray techno-scientific knowledge as being much closer to the everyday commonsense experience rather than to the body of experts' knowledge" (Dimopoulos, Koulaidis, and Sklaveniti 2003: 205). Where scientific images are abstract, schematic, conventionalized and coded, the photograph is concrete and detailed, presenting itself "as a naturalistic, unmediated, uncoded representation

5. The use of the words *real* and *realistic* as opposed to *scientific* is debatable in the context of scientific communication. Kress and van Leeuwen (2006: 158) explain that whether diagrams are seen as less or more 'real' than photographs will depend on definitions of reality, on values, beliefs and interests of each social group: "reality may be in the eye of the beholder, but the eye has had a cultural training, and is located in a social setting and a history."

of reality" (Kress and van Leeuwen 2006: 157–158). In between these two extremes of the continuum are images that result from scientific imaging techniques that compensate for human vision deficits, such as microscope images. These techniques can be seen as processes "of 'lending' invisible objects or phenomena the optical properties of familiar objects – making them fit in to our experience of sight" (Nieman 2000: 57).

Adapting the categories used by Hullman and Bach (2018) and Yoon and Chung (2017), we found five types of images in the GAs, from less to more abstract: *illustration*, *diagram*, *data visualization*, *scientific visualization*, and *symbolic notation* (Table 3). A total of thirty-eight instances of these types were identified: in GAs that presented two visual entities of the same type, the image was counted once (*symbolic notation* in CGA#10, Table 1); if the GAs showed two visual entities of different types, each entity was coded according to the type of image it represents (in Table 1, *illustration* and *data visualization* in CEGA#24, and *scientific visualization* and *data visualization* CEGA#29). In GAs where several visual entities form a single *diagram*, it was counted as one instance of this kind of image (CEGA#21 in Table 1, where *illustration* and *symbolic notation* are combined into a circular diagram).

Table 3. The nature of images in GAs

	Nature of images (LESS SPECIALIZED ← → MORE SPECIALIZED)									
	Illustration		**Diagram**		**Data visualiz.**		**Scientific visualiz.**		**Symbolic notation**	
	n	%	n	%	n	%	n	%	n	%
Review GA	2	25	4	50	1	12.5	–	–	1	12.5
Experim. GA	5	16.6	3	10	10	33.3	7	23.3	5	16.6
Totals	7	**18.4**	7	**18.4**	11	**29**	7	**18.4**	**6**	**15.8**

Our results show that while abstract images predominate (24 instances, 63.1%, *data visualization*, *scientific visualization*, and *symbolic notation*), more realistic images in the form of *illustrations* and *diagrams* including illustrations (14 instances, 36.8%) also play a significant role in bringing a lower level of *formality* to the genre. *Illustrations* can be drawings and cartoons that resemble "vernacular ways of expression or lying very close to the realistic appearances of things" (Dimopoulos, Koulaidis, and Sklaveniti 2003: 191). They somehow replicate real-world elements or ideas, but with lower naturalism than in photographs. In our sample, less prominent *illustrations* were small images showing (parts of) people, glands, cells, and symbolic things (e.g. lightning, drop of water, skull and crossbones). Both review GAs (2, 25%) as well as experimental GAs (5, 16.6%) use *illustrations*.

Only one GA included *photos* within a diagram, similar to the result found by Hullman and Bach (2018), with rare use of *photos* in GAs, but expressive use of *illustrations* as the predominant type of image. Considering that these authors included microscope images under the category *photos*, it is uncertain whether any actual photograph as those taken with an ordinary camera were present in their sample. However, in social sciences GAs, Yoon and Chung (2017: 1375) found that traditional camera photographs were the third most recurrent type of image, after *illustrations*.

In spite of the differences in patterns of photograph usage in GAs found in the previous studies and in our work, probably due to disciplinary cultures, the considerable recurrence of *illustrations* in GAs can be linked to the potential of GAs in increasingly attracting readers from outside the expert fields of the journals, including a non-specialized audience, but also readers from interdisciplinary fields and/or from related fields of expertise.

The editorial policies of seven journals may be interpreted in this sense, as they indicate that the GA should "capture the attention of a *wide readership online*" (*Acta Biomateralia, Acta Materalia, Insect Biochemistry and Molecular Biology, Scripta Materialia, Powder Technology, Applied Catalysis B: Environmental*), or that it "should capture the eye and curiosity of a *broad spectrum of readers*" (*Chemical Reviews*). The authors in our study, however, do not see this as a motivation for the publication of GAs. They agree that one of the reasons why GAs were introduced in scientific papers is to attract readers, but not necessarily *lay* readers.

Nonetheless, the predominant type of image that is encouraged by editorial policies in chemistry and engineering is the typical abstract scientific image. The documents explicitly regard photographs of people, cartoons, clip art and smiley faces as unacceptable. Appropriate images are those that "reduce distracting and cluttering elements" (*Acta Biomaterialia*), are "closely tied to the science in the article" (*Journal of Physical Chemistry Letters*) and "should uphold the standards of a scholarly, professional publication" (*Organic Letters*). The GAs in our sample revealed a predominance of abstract images in the form of charts and chemical schemes.

We found no significant difference in the nature of images used in the three areas. In relation to the types of GAs (review and experimental), we noted that *diagrams* were more frequently used in GAs of review articles (4 GAs, 50%), while *data visualization* predominated in experimental GAs (10 GAs, 33.3%). *Diagrams* "employ various common symbols and spatial mappings to express ideas, concepts, and processes" (Hullman and Bach 2018: 12), "using standardized shapes, such as lines, boxes, circles, arrows" (Yoon and Chung 2017: 1373). *Data visualization* includes representations conventionally associated to science that quantitatively or qualitatively pack large amounts of data such as charts, tables, networks, heat maps (Yoon and Chung 2017; Hullman and Bach 2018). *Diagrams* were also found to be the dominant type of image in GAs from the social sciences (Yoon and Chung 2017: 1375), and in Hullman and Bach (2018: 12), it was the second most frequent type of visual representation, after illustrations. We also found that only experimental GAs present *scientific visualizations* (a total of 7, 23.3%). This kind of scientific images "are intended to faithfully represent these objects to allow exploration and analysis of their structures; for example, brain tumors or functional MRI data" (Hullman and Bach 2018: 11), for example, EGA#18 (left) and CEGA#29 (left) in Table 1. Finally, six GAs (15.8%) use highly coded types of images called *symbolic notation*, which consist of "graphical codes that use mostly domain specific symbols and compositions" (Hullman and Bach 2018: 11) such as chemical schemes (CGA#10 and CEGA#21 in Table 1) and mathematical formulae.

In spite of the predominance of abstract images in our sample, we noted that their formality was frequently softened by the use of color. It was actually used in all types of images, in twenty-seven GAs (90%). In some texts this use is subtle, limited to small units (such as the blue O and red N in CGA#10 and the arrows and the verbal label "graphene encapsulation" in EGA#18 – Table 1) or to a single color. In most GAs, however, color use is rich in variety and saturation, with eleven GAs (36.6%) using more than eight different colors simultaneously, including primary blue, red and yellow, bright orange, neon green, vivid cyan blue. Six GAs (20%) also use a background color, which may be bright as in CEGA#21 (Table 1).

The use of color for aesthetic purposes is typical of contexts such as Art and Media as a "source of pleasure and affective meanings" (Kress and van Leeuwen 2006: 165). It has a promotional role as it "can increase the reader's attention span by more than 80 percent" (Kress and van Leeuwen 2002: 349). However, in the context of science, the use of color has only recently become more frequent, influenced by contemporary technological affordances for the production of scientific images (access to software) and distribution of journals online. Such uses of color have been justified only when they work as a tool to depict scientifically relevant data, not as a distraction. Higher degrees of abstraction in science use black and white as the canonical feature of scientific images. Even when color is added to a microscope image of a cell in order to highlight important information, it will be a plausible color. When instead "bright and eye-catching" colors are used, "they signal clearly that they are 'false colors' and are chosen either to make an object of interest stand out or to transform an image of visually dull objects such as single cells into a striking image in its own right" (Nieman 2000: 93). This kind of "pretty pictures" are "totally unrelated to science," but play a major role for promoting research (Nieman 2000: 93).

The journal policies in our sample encourage the use of color in GAs. Only *Nano Letters* limits it "to those instances *where it is needed for clarity*." The majority of the participants (75%) agree that color is essential in the GA. This feature could be related to the advertising purpose of the GAs and work as evidence that the genre is not exclusively directed to expert readers. Nevertheless, interpretations in this sense are highly speculative because color is also extensively used inside 25 (83.3%) of the articles.

6. Final remarks

What then is the relationship between traditional verbal abstracts, images in research articles and GAs? We see this relation as complementary, rather than overlapping. The GA mixes the purposes of verbal abstracts with the 'lexicogrammar' of images in research articles. Like the verbal abstract, it anticipates content from the research article, but complements the verbal abstract by allowing the use of images to do so. The value of images in science puts the GA into a more powerful position than verbal abstracts to attract readers in the online competitive space. We therefore consider the GA a new digital genre in its own right with *bridging* properties that "unite the best of two worlds" (Herring et al. 2005: 162) to fulfill new purposes.

Nevertheless, our results suggest that the new purposes and form of GAs are under construction. Evidence for this can be found in the lack of uniformity in the genre in terms of layout, originality, and the nature of images within and across disciplines as well as in the divergences among editorial policies about the purpose and form of GAs. Previous work on GAs in a variety of fields (Lane, Karatsolis, and Bui 2015; Yoon and Chung 2017; Hullman and Bach 2018) found similar diversity. We observe some homogeneity in the perceptions of the (co)authors about GAs, but due to low participation rates, any generalizations would be premature.

The variety of possibilities for presenting GAs can be linked to the novelty of the genre still in an evolutionary stage in which most effective practices are under construction. While the *what* (visual summary) and the *why* (to attract readers) of GAs are less obscure, the *how* is still remarkably vague (Cox 2015). The participants in our study confirm that the instructions for authors were hardly helpful in elaborating GAs and that they could not find information elsewhere to help them. Our study reinforces the conclusion of previous studies about the need for further investigations of the GA genre in different disciplines complemented with insiders' views and practices. We also found evidence of commodification forces in the cosmetic use of color and in drawings of everyday things that non-expert readers can relate to. Both these features can be interpreted as the colonizing of academic discourse by media discourse for the promotion of journal publications. The knowledge derived from these studies could help in the development of pedagogical actions to promote multiliteracies for writing and reading GAs.

References

Askehave, Inger, and John M. Swales. 2001. "Genre Identification and Communicative Purpose: A Problem and a Possible Solution." *Applied Linguistics* 22 (2): 195–212. https://doi.org/10.1093/applin/22.2.195

Ayers, Gael. 2008. "The Evolutionary Nature of Genre: An Investigation of the Short Texts Accompanying Research Articles in the Scientific Journal *Nature*." *English for Specific Purposes* 27: 22–41. https://doi.org/10.1016/j.esp.2007.06.002

Bezemer, Jeff, and Gunther Kress. 2008. "Writing in Multimodal Texts: A Social Semiotic Account of Designs for Learning." *Written Communication* 25 (2): 166–195. https://doi.org/10.1177/0741088307313177

Breeze, Ruth. 2016. "Tracing the Development of an Emergent Part-Genre: The Author Summary." *English for Specific Purposes* 42: 50–65. https://doi.org/10.1016/j.esp.2015.11.003

Cox, Luk. 2015. *Are Graphical Abstracts Changing the Way we Publish?*. Last accessed June 15, 2019. https://www.wiley.com/network/researchers/promoting-your-article/are-graphical-abstracts-changing-the-way-we-publish

Devitt, Amy J. 2004. *Writing Genres*. Carbondale, IL: Southern Illinois University Press.

Dimopoulos, Kostas, Vasilis Koulaidis, and Spyridoula Sklaveniti. 2003. "Towards an Analysis of Visual Images in School Science Textbooks and Press Articles about Science and Technology." *Research in Science Education* 33: 189–216. https://doi.org/10.1023/A:1025006310503

Elsevier. 2012. *The Article of the Future*. Last accessed June 15, 2019. https://www.elsevier.com/connect/the-article-of-the-future

Fairclough, Norman. 1992. *Discourse and Social Change*. Cambridge: Polity Press.

Fairclough, Norman. 1995. *Critical Discourse Analysis*. London: Longman.

Gilaberte, Yolanda, Eduardo Nagore, Salvador Arias-Santiago, and David Moreno-Ramirez. 2016. "¿Vale una Imagen más que Mil Palabras?: La «Figura Resumen» o *Graphical Abstract*»." *Actas Dermosifiliográficas* 107 (7): 545–546. https://doi.org/10.1016/j.ad.2016.07.001

Gu, Xin, and Karen Blackmore. 2016. "Recent Trends in Academic Journal Growth." *Scientometrics* 08 (2): 693–716. https://doi.org/10.1007/s11192-016-1985-3

Hemais, Barbara J. W. 2014. "Word and Image in Academic Writing: A study of Verbal and Visual Meanings in Marketing Articles." *ESP Today* 2 (2): 113–133.

Herring, Susan C., Lois A. Scheidt, Elijah Wright, and Sabrina Bonus. 2005. "Weblogs as a Bridging Genre." *Information Technology & People* 18 (2): 142–171. https://doi.org/10.1108/09593840510601513

Huckin, Thomas. 2001. "Abstracting from Abstracts" In *Academic Writing in Context: Implications and Applications*, ed. by Martin Hewings, 93–103. Birmingham: The University of Birmingham Press.

Hullman, Jessica, and Benjamin Bach. 2018. "Picturing Science: Design Patterns in Graphical Abstracts." In *Proceedings of the 10th International Conference on the Theory and Application of Diagrams (Diagrams)*, 1–16. Edinburgh, UK.

Hyland, Ken. 2009. *Academic Discourse: English in a Global Context*. London/New York: Continuum.

Johns, Ann M. 1998. "The Visual and the Verbal: A Case Study in Macroeconomics." *English for Specific Purposes* 17 (2): 183–197. https://doi.org/10.1016/S0889-4906(97)00010-0

Kress, Gunther, and Theo van Leeuwen. 2002. "Colour as a Semiotic Mode: Notes for a Grammar of Colour." *Visual Communication* 1 (3): 343–368. https://doi.org/10.1177/147035720200100306

Kress, Gunther, and Theo van Leeuwen. 2006. *Reading Images: The Grammar of Visual Design*. London: Routledge. https://doi.org/10.4324/9780203619728

Lane, Suzanne, Andreas Karatsolis, and Lily Bui. 2015. "Graphical Abstracts: A Taxonomy and Critique of an Emerging Genre." In *Proceedings of the 33rd Annual International Conference on the Design of Communication*. Limerick, Ireland. https://doi.org/10.1145/2775441.2775465

Lowe, Derek. 2012. *Graphical Abstract Tedium*. Last accessed June 15, 2019. http://blogs.sciencemag.org/pipeline/archives/2012/09/06/graphical_abstract_tedium

Miller, Carolyn R., and Dawn Shepherd. 2004. *Blogging as Social Action: A Genre Analysis of the Weblog*. University of Minnesota. Retrieved from the University of Minnesota Digital Conservancy. Last accessed June 15, 2019. http://hdl.handle.net/11299/172818

Miller, Thomas. 1998. "Visual Persuasion: A Comparison of Visuals in Academic Texts and the Popular Press." *English for Specific Purposes* 17 (1): 29–46. https://doi.org/10.1016/S0889-4906(97)00029-X

Motta-Roth, Désirée, and Viviane M. Heberle. 2015. "A Short Cartography of Genre Studies in Brazil." *Journal of English for Academic Purposes* 19: 22–31. https://doi.org/10.1016/j.jeap.2015.05.006

Nature Chemistry. 2011. "Editorial: The Art of Abstracts." *Nature Chemistry* 3: 571. https://doi.org/10.1038/nchem.1109

Nieman, Adam. 2000. The Popularisation of Physics: Boundaries of Authority and the Visual Culture of Science. Unpublished thesis. Faculty of Applied Sciences and Faculty of Humanities, University of the West of England, Bristol.

Painter, Clare, James R. Martin, and Len Unsworth. 2012. *Reading Visual Narratives: Image Analysis in Children's Picture Books*. Sheffield, UK; Bristol, USA: Equinox.

Pérez-Llantada, Carmen. 2012. *Scientific Discourse and the Rhetoric of Globalization. The Impact of Culture and Language*. London/New York: Continuum.

Pérez-Llantada, Carmen. 2013. "The Article of the Future: Strategies for Genre Stability and Change." *English for Specific Purposes* 32: 221–235. https://doi.org/10.1016/j.esp.2013.06.004

Swales, John M. 1990. *Genre Analysis: English in Academic and Research Settings*. Cambridge: Cambridge University Press.

Swales, John M. 2004. *Research Genres: Explorations and Applications*. Cambridge: Cambridge University Press. https://doi.org/10.1017/CBO9781139524827

Swales, John M., and Christine B. Feak. 2009. *Abstracts and the Writing of Abstracts*. Ann Arbor, Michigan: University of Michigan Press. https://doi.org/10.3998/mpub.309332

Todorov, Tzvetan. 1976. "The Origin of Genres." *New Literacy History*, 6: 159–170. https://doi.org/10.2307/468619

Yoon, JungWon, and EunKyung Chung. 2017. "An Investigation on Graphical Abstracts Use in Scholarly Articles." *International Journal of Information Management* 37 (1): 1371–1379. https://doi.org/10.1016/j.ijinfomgt.2016.09.005

CHAPTER 5

Scholarly soundbites

Audiovisual innovations in digital science and their implications for genre evolution

Elizabeth Rowley-Jolivet and Shirley Carter-Thomas
Laboratoire Ligérien de Linguistique (UMR 7270, Université Orléans-Tours, BnF, CNRS) / IMT-BS and LATTICE (CNRS, ENS & Université Paris 3; PSL & USPC)

This study investigates a recent web-enabled feature, the use of brief audio/video recordings for the communication of scientific research findings to a non-specialized audience, and discusses the implications of these "scholarly soundbites" for genre evolution in the digital environment and for the mediatization of science. We focus on four types of audiovisual material, all characterized by their brevity: Three-Minute Thesis presentations, author videos, and podcasts on a popular science and a research journal website. An analysis of the moves and of the recontextualization strategies used to manage the knowledge asymmetry between scientists and audience highlights differences between the four types of soundbites as well as with the corresponding written genres (research articles, PhD dissertations).

Keywords: scholarly soundbite, Three-Minute Thesis, science podcast, author video, digital science, genre evolution, move analysis, recontextualization, mediatization of science

1. Introduction

Science researchers speak with different voices in different contexts and genres, as has been amply demonstrated in previous work by sociologists of science and discourse and media analysts. The pioneering studies in the early 1980s by sociologists (e.g. Knorr-Cetina 1981; Gilbert and Mulkay 1984) into lab work and talk and informal oral exchanges among scientists painted a picture of scientific discourse behind the scenes that was very different from that of the public face of science. A second wave of studies, this time among linguists and discourse

https://doi.org/10.1075/pbns.308.05row
© 2019 John Benjamins Publishing Company

analysts, starting in the late 1990s focused on established scholarly speech genres such as the conference presentation or thesis defense as a necessary corrective to the almost exclusive focus on the written tenure genres of the journal article or monograph (Ventola, Shalom, and Thompson 2002; Swales 2004). As all this work has made clear, scientists "do not confine themselves to the kind of language used in published articles; they move between several repertoires" (Myers 2003: 270): the mode (spoken or written), the intended audience (a semi-private or a public record), and the stage in the research process documented (exploratory in the lab, often ongoing in the conference talk, after completion of the research in the journal article) are all major sources of differences that impact both how the researchers express themselves and what aspects of their work they focus on.

We are now witnessing what could be called a third wave, as the dissemination of science via the Internet has immeasurably enlarged the variety of repertoires, contexts, and genres, in which scientists can present and discuss their work. The Internet has changed the game rules for scientific communication in several respects: the need to address a global, indeterminate audience, not an esoteric circle of peers, raises the problem of knowledge asymmetry between science researcher and audience, and requires suitable discursive strategies; the digital medium is characterized by heightened interactivity and is an expressively free, often strongly personalized, environment; it is extremely reactive timewise but may, critically, shorten the attentional timespan of users, accustomed to zapping from one page or site to another and to processing small bite-sized pieces of information (see similar claims in Orpin this volume; Smart and Falconer this volume; Reid and Anson this volume). It may also lead to the increased introduction of promotional features into established genres (e.g. Bhatia 2005). Lastly, the Internet revolution has involved a shift from primarily print media to audiovisual material adapted to mobile devices.

In response to the new social needs generated by this upheaval, scholarly journals have increased their digital presence by uploading existing genres unchanged to the Internet, or adapting them in various ways to the hypertextual and multimodal affordances of the medium (see Pérez-Llantada 2013, 2016; Harmon this volume), while new hybrids and web-native genres, developed both by journals and by the scientists themselves, have emerged – scholarly blogs, scientific wikis, video clips, podcasts, TED talks, open science notebooks to mention but a few (Myers 2010; Campagna et al. 2012; Carter-Thomas and Rowley-Jolivet 2017), making it increasingly difficult to maintain the traditional distinctions between academic genres and questioning or blurring the boundaries between scholarly and popular science (Myers 2003).

In addition to these factors related to the Internet medium itself, a deeper societal process that may also induce genre evolution is mediatization. Starting from the recognition that digital media have become ubiquitous in time (available 24/7), space (the audience is global) and context (pervasive in all areas of social life), media scholars argue that these quantitative extensions of the media landscape have led to qualitative changes in how other social subsystems and institutions (politics, education, religion, science…), which each had their own traditional modes of operation, values and priorities, now communicate and operate. The media logics of marketization, popularization, immediacy, and the pursuit of publicity and visibility are tending to permeate or even become embedded in how these institutions function. While this may be most obvious in the political sphere, where the news agenda is increasingly set by what is trending on Twitter (Garland, Tambini, and Couldry 2018), academia and science are also affected: the worldwide university ranking systems turn universities into competitive providers on an educational market (Pallas and Wedlin 2013), academic networks such as ResearchGate turn researchers into "entrepreneurs of themselves" in a marketplace of ideas (Hammarfelt, de Rijcke, and Rushforth 2016), while the major academic publishers have adopted aggressive marketing strategies online and compete for high impact factors.

In the present study, we will attempt to interpret our results both in terms of mediation (the role of the Internet medium in genre evolution) and in terms of mediatization, pointing out whether any signs of these new media logics are discernible in the communication of science online in our data. We have chosen to focus on oral data, relatively under-researched compared to written web-mediated documents, in view of their prevalence among Internet users. We collected a corpus of four types of short audio or video recordings from the Internet, in which junior and senior researchers talk about their ongoing or recently published research: Three-Minute Theses (henceforth 3MTs), podcasts from the popular science magazine *Scientific American*, podcasts from the research journal *Nature*, and author videos from the *Nature* Video Channel or Elsevier journal websites. We have called these recordings "scholarly soundbites" as they are all characterized by their brevity. 3MTs, by definition, never exceed 180 seconds, whereas, as proclaimed on the website of the University of Queensland (Australia), the creator of the 3MT competition, "An 80,000 word PhD thesis would take 9 hours to present. Their time limit… 3 minutes."[1] The *Scientific American* podcasts are broadcast in the feature called "60-second Science," and while the author videos and *Nature* podcasts are more variable in length, ranging from 1 min 30 s to 6 minutes, they are very brief in terms of word count compared to the corresponding research article (RA). Although we have not yet reached the stage of "Twitter science," we can wonder whether the scholarly soundbites analyzed here are not a step in this direction.

1. <https://threeminutethesis.uq.edu.au>

These four scholarly soundbites are not to be considered representative of a single genre as they cover a fairly wide range of discourses. Our aim was to attempt an overview of the palette of spoken resources available on Internet for scientific research communication to a predominantly non-specialized audience. After presenting the data collected (Section 2), we first undertake a move analysis of the soundbites, inquiring into how the scientific content is adapted in comparison, where appropriate, with their written equivalents (theses and RAs). In Section 4 we then analyze the discursive strategies used to recontextualize the research, focusing on four aspects that reflect the adaptations required by the context and/or the digital medium: reformulation and repetition strategies that have both an explanatory and a persuasive function (4.1); illustration procedures such as the use of comparison and metaphor to make the content understandable and attractive for a non-specialist audience (4.2); interactivity through the use of questions (4.3); and personalization through researchers' comments on their work (4.4). The concluding section attempts to draw out some of the implications of the changing ways science is communicated in the digital age.

2. Data description

As shown in Table 1, in each of the four datasets, we collected fifteen recordings from a range of sciences over the period 2010–2018. All the documents were transcribed, or if transcriptions were already available on the websites (the case of the podcasts), we checked the transcripts and made some corrections. The total length of the corpus comes to 229 minutes, and the total word count to just over 40,000 words. The average length of the four types of soundbites is, in increasing order, about 2 min (*SciAm* Podcasts), 3 min (3MTs), 4 min (author videos) and 6 min (*Nature* podcasts). The podcasts are audio recordings only, whereas the 3MTs and author videos were video-recorded, enabling the inclusion, in addition to sound effects, of some visual material. The number and type of speakers vary depending on the category. As we wished to investigate how the research scientists themselves speak about their science in digital media, we excluded recordings in which only a science journalist or a voice-off commentary could be heard, as well as those that did not refer to a recent publication but were an overview or introduction to a topic (the case of many science videos available online). For the podcasts, starting from January 2017, we selected the first 15 recordings that met these criteria; for 3MTs, given the time-lag of several years between the 3MT presentation and publication of the thesis, we had to go further back in time, and selected only those where the thesis was available; for the videos, we selected all the science videos on the *Nature* video channel that met our criteria, and completed the dataset with 5 from Elsevier journals (see below).

Table 1. The audiovisual corpus

	3-Minute Theses	Author Videos	*SciAm* Podcasts	*Nature* Podcasts	Total
Medium	video	video	audio	audio	
Number	15	15	15	15	60
Total length (minutes)	**43′50″**	**62′38″**	**29′5″**	**94′13″**	**229′46″**
Average length (minutes)	**3′**	**4′15″**	**2′**	**6′20″**	
Total words	**7,337**	**9,441**	**4,509**	**18,956**	**40,243**
Average words	**489**	**630**	**300**	**1264**	
Date	2011–15	2010–18	2017	2017	
Speakers*	1 PhD student	1S Or (2 cases) 1S + 1J	1J + 1S	1 or 2J + 1 or 2S	

(*) J: journalist; S: scientist

3MT presentations are a new, competitive academic speech genre that gives PhD students the opportunity to present in three minutes their ongoing research to a mixed disciplinary audience before a panel of judges. The 3MT™ competition itself, we would contend, can be seen as evidence of the marketization of academic research in that 3MT is the registered trademark of a higher education institution, the University of Queensland, which defines the rules and authorizes other universities to use the "brand," and has contributed greatly to enhancing the media profile of this institution. The contest applies the typical elements of game-playing – scoring with winners and losers, competition with others, strict rules of play – and can be considered an example of the "gamification of science" (Hammarfelt, de Rijcke, and Rushforth 2016). Presentations are limited to three minutes, the presentation must be spoken word only, with no props and a single static slide. The presenters are novice researchers who deliver a monologue but are not required to answer any questions at the end. The adjudicating panel attributes the awards, and the judging criteria are clearly laid out: candidates should "avoid jargon and academic language," and clearly describe the topic, significance, results and outcomes of the research; they are advised to "tell a story," to be enthusiastic and to use an opener that catches the audience's attention. In contrast, the other three categories all involve established researchers and are positioned at a later stage in the knowledge production process as they discuss recently completed and published research findings.

The fifteen author videos were downloaded from two sources: ten from the *Nature* Video Channel[2] and five attached as Featured author videos to RAs

2. https://www.youtube.com/channel/

published in Elsevier journals.[3] Few researchers appear to have availed themselves of this possibility, however, apparently because of the lack of established guidelines or genre conventions, according to some researchers we consulted. Our data reflect this, as unlike the 3MTs, the length of the videos is variable (from < 2 to 6 minutes) as are the number and type of speakers: a regular alternation of turn-taking between scientist and journalist in two, a very brief introduction by the journalist in another two, and the scientist(s) as sole speaker(s) in the remaining eleven.

The remaining two categories, *Scientific American* podcasts and *Nature* podcasts are audio recordings only. The popular magazine and the research journal broadcast podcasts every weekday and once a week, respectively. This initiative can be seen as an indication of the mediatization of the scientific press, both in the accelerated timeframe of some of the podcasts (daily updates, by analogy with news media) and in the increased visibility the podcasts confer on the journals. In contrast to the two video categories, journalists play an important role. In the *SciAm* podcasts, their role is even the dominant one: the recently published research is introduced, summarized and commented on by the journalist, and the scientist's role is restricted to a brief quote inserted into this commentary – literally, a 'soundbite' of a few seconds. There is no direct dialogue between the journalist and scientist, and the podcasts are very brief.

The *Nature* podcasts are the longest of our four types of recordings, averaging six minutes, which allows more room to develop the science and to give the scientists a more prominent role, and can therefore be assumed to address a more scientifically literate – or at least interested – audience than the *SciAm* podcasts. Here, there is a real exchange between the journalist and scientist (see Section 4). In some cases, the publication of the article is used as the launchpad for a broader coverage of the topic, bringing in previous work on the subject or competing hypotheses, leading to multiple voices.

We also collected the corresponding written texts: the 15 thesis Abstracts and the theses themselves for the 3MTs, and the 45 RAs for the other three categories. The PhD Abstracts sub-corpus contains 7,123 words and the theses cover 3,320 pages. The forty-five RAs total 200,430 words.

3. Move analysis

Move analysis is a frequently used tool in genre analysis to determine the recurrent rhetorical structure of a given discourse (Swales 1990). We applied move analysis to investigate which aspects of the scientific content predominate in these new

3. https://www.elsevier.com/authors/author-services/featured-author-videos

modes of spoken scholarly communication, addressed to a non-specialist audience. To enable comparison between our four subsets, we adopted the same move categories across the board. Table 2 summarizes the eight main moves occurring in the soundbites.

Table 2. Move structure in the scholarly soundbites*

Move	3MTs	Author Videos	*SciAm* Podcasts	*Nature* Podcasts
1 Orientation/Introduction	100	67	87	100
2 Rationale & Purpose	97	67	73	80
3 Methods	94	87	80	67
4 Results	73	100	100	100
5 Discussion of Results	0	33	0	80
6 Implications/Applications	87	73	87	73
7 Rounding off	100	0	67	100
8 Reference to publication	0	33	100	100

*Expressed as % of talks or recordings in which the move occurs (n = 15 in each subset)

Move 1, Orientation or Introducing the Topic, consists generally of a short preamble preparing the audience for the topic of the talk. Move 2, Rationale and Purpose, explains the motivation for and the aim of the research, corresponding to the "Establishing a niche" and "Occupying the niche" moves in Swales' model for research introductions (1990, 2004), grouped together here as in these brief talks to a non-academic audience the two were often conflated. Moves 3, 4, 5, and 6 likewise reflect the IMRD development of the RA. In Move 7, Rounding off, the speaker indicates the close of the talk or recording using various strategies, discussed below. Move 8, which only occurs systematically in the podcasts, is a set formula referring the listener to the corresponding article; the transcripts on the website give the hyperlink to the article at this point. The apparent similarity of most of these moves to the traditional rhetorical structure of the RA is unsurprising since all the soundbites are accounts of scientific research, but there are several significant differences among the four subsets, and some interesting characteristics that differentiate them from the corresponding written data. For reasons of space, only some of the main points will be highlighted below.

Among the strategies used in Move 1 to introduce the topic, for example, the audience is often addressed directly, either by a statement:

(1) J: You've probably seen pictures of Greek villages, where every house is painted bright white. (SciAm)

or by a question (see 4.3). Alternatively, the creation of an imaginary scenario (see 4.2) or banter between two journalists, including puns and homely examples, leads listeners gradually into the scientific topic:

(2) J1: Adam, open wide please. I need to look at your dental calculus.
J2: My what?
J1: Your dental calculus: calcified dental plaque. Basically, bits of food and microbes that live among your pearly whites.
J2: No, I'd really rather you didn't. My teeth are spotless and you're definitely not a dentist.
J1: Correct! But I am very interested in what dental plaque can reveal about diet and health. A new study in *Nature* takes a look at some dental calculus from Neanderthal teeth... (*Nature*)

These engaging and interactional openings clearly serve to 'hook' the listener by relating the topic to his/her own experience (holidays abroad, visits to the dentist).

Two of the conventional sections of the RA, Theoretical Background and Methods, are handled very differently in the soundbites. Theoretical developments are quite simply absent. We found only one occurrence of Hu and Liu's (2018) Theoretical Framework move in our own 3MTs and one other in the author videos, both in mathematical physics where theoretical investigation was in fact the purpose of the research, and none in the thirty podcasts. We therefore excluded this move from our analysis. Although Move 3 (Methods) is frequent in our spoken data, the information given about the methods is generally extremely brief. A typical example is (3), where the fifteen pages of Method and Supplementary data in the RA are reduced to a single sentence in the corresponding podcast:

(3) J: They popped off bits of ancient dental plaque and then sequenced the DNA contained within. (SciAm)

A salient difference among the four subsets concerns the Results and Discussion of Results moves. The videos and podcasts, designed to illustrate and promote a recent publication, all contain Move 4, Results: in the two shortest videos this constitutes practically the only move, and also receives great prominence in the *SciAm* podcasts (given twice in six cases). In the 3MTs, in contrast, results feature in only three-quarters, confirming Hu and Liu's (2018) data where Results occurred in only 57% of 3MTs, and was therefore considered an optional move applying a cutoff point of 80%. This can be attributed to the ongoing nature of the thesis research, as can the total absence of Move 5, Discussion of the results. The latter is also absent from the *SciAm* podcasts, despite the fact that the podcast is based on an RA where the Discussion section is a crucial feature. We attribute this to the extreme brevity of these podcasts and the popular audience targeted. These snippets give only a superficial, factual view of scientific research.

In the *Nature* podcasts, in contrast, the Discussion move can be considered an obligatory one. The 'facticity' of scientific results (Latour 1987), which seems to be taken for granted in the *SciAm* podcasts, is here problematized as either the scientists themselves discuss possible interpretations of their findings, or, more agonistically, their claims are contested by others. The podcast becomes a forum for the confrontation of various viewpoints and hypotheses:

(4) S: We have people in the field that are sceptical and people in the field that are lauding our accomplishment, and it turns out that the people in the field that are sceptical are competitors of ours, and they've made comments like, "I don't believe that you can get that high of a pressure." Fine, we've shown what we've done. So they can go ahead and try it. If they try it, they're either going to confirm it or show that it didn't happen. I'm confident that they'll find that it happens. (*Nature*)

In the videos, this move is less frequent (occurring in 5 videos), but nonetheless is an interesting exploitation of the digital medium in that, as in some *Nature* podcasts, the two conflicting hypotheses are embodied by two different researchers, creating a multiplicity of voices.

The closing move, Rounding Off, often uses similar strategies to the opening move. 3MT presenters invariably indicate the end of their talks with a short "thanks" and a smile to the audience, and some with a punchline or an echo strategy, while in the two podcast sets the journalist often concludes on a light-hearted note by punning on the keywords of the topic (see Section 4.1). This move is absent in the Videos, however, which often end abruptly.

Move structure is therefore followed and developed to varying degrees in the four subsets. In the 3MTs, the constraining rules and recommendations of the contest induce considerable uniformity in rhetorical structure. Although certain moves may occur recursively, we found that the ordering of the moves is remarkably regular: moves 1, 6 and 7, due to their very nature, always occupy the same slots, and intervening moves also follow the linear order in the overwhelming majority of cases. The move structure of the *SciAm* podcasts is also highly formatted, with the same linear move order in almost all the broadcasts.

In the *Nature* podcasts the move structure is more fully developed than in the other three subsets, as all eight moves occur frequently, six in over 80% of the podcasts, and the other two (Method and Implications/Applications) in two-thirds of them. The structure is, however, fairly flexible, progressing by a question and answer format (see Section 4.3).

A move analysis gives a misleading picture of the videos, however, as no recurrent rhetorical structure could be detected: the videos vary considerably in the order of the moves, in their number (ranging from 1 to 7, average 3.5), and in the time devoted to each in any given video. Only two moves can be considered obligatory (Methods and Results). There also appears to be a great variability in the type of audience targeted. Some of the *Nature* videos lean towards the popular science end of the spectrum, exploiting the visual impact of the medium by showing striking films, including music or sound effects, or are descriptive rather than analytical. At the other extreme, some of the Elsevier videos targeting a highly specialized audience resemble the abstract of the corresponding RA in content and terminology. The scientists do not seem to have received any guidance, however, on whether their research topic was suitable for a video presentation or not: some of the videos are simply 'talking heads', whereas others are greatly enhanced thanks to the visual medium.

4. Recontextualization strategies in the soundbites

In the linear diffusion or deficit model of science communication, knowledge transfer is seen as a one-way process from scientist to (semi)-lay public via a mediator, usually a journalist, and a dumbing down of the original source. In preference to this reductionist view, we adopt here the approach of several scholars (e.g. Jacobi 1999; Ciapuscio 2003; Gülich 2003) who see expert-lay communication as a *recontextualization* of the initial source for different addressees, which can be accomplished either by the scientists themselves or through collaborative 'work' with journalists. In this process, several types of discursive strategies can be used. Selecting the most appropriate formulation is obviously crucial. It is also often necessary to illustrate the concepts and findings by procedures such as comparisons and analogy, exemplification and metaphor, in order to relate the knowledge to the receiver's own context, or to 'humanize' the impersonal scholarly discourse by bringing in the researcher as a person. Additionally, as success in oral communication is an interactional achievement (Thompson 1998), interactive devices such as questions are a useful strategy.

This section therefore first focuses on the reformulation procedures used to make the scientific content accessible to the lay audience (4.1). We then look at a closely related set of discursive devices used for comparison and exemplification that we have grouped under the label illustration procedures (4.2), and follow with an analysis of the role of questions (4.3). The section concludes (4.4) by highlighting the human side to science through the expression of scientists' comments on their work. In the *Nature* and *Scientific American* podcasts, we will also distinguish between journalists' and research scientists' use of these strategies, as this sheds light on their respective roles.

4.1 Reformulation and repetition strategies

By reformulating, speakers can rework or "treat" again (Gülich 2003) an earlier segment of talk – whether produced by themselves or by another interlocutor – in order to better adapt it to the communicative context. Previous work has often focused on a selection of specific reformulation markers: *that is (to say), i.e., in other words, namely, I mean* (Pennec 2006). However, an initial search for markers of this type in our spoken corpus revealed only one occurrence of *that is (to say)* over the four data sets. We therefore adopted an onomasiological and discursive approach, involving a close reading of reformulation and repetition patterns in our corpus, and distinguished four types of procedure:

1. Deliberate repetitions. Discarding examples where the repetition seemed due to some type of disfluency, we only considered cases where the repetition seems deliberately adopted for stylistic or rhetorical reasons.
2. Paraphrastic reformulations. The reformulation is a formal variation of the original expression; the content is not changed or only very marginally.
3. Paraphrastic reformulations involving definitions or acronyms. This category is the same as (2) but, given the highly specialized nature of our corpus, we wanted to see to what extent the paraphrases specifically concerned definitions.
4. Reformulation with specification/expansion. The reformulation here involves a variation in both form and content, with information being added, expanded or analyzed from a different angle. Table 3 gives the occurrences in the data.

Table 3. Reformulation and repetition strategies

Type of reformulation	**3MTs**	**Author videos**	**SciAm podcasts**	***Nature* podcasts**	***Total***
1 Deliberate repetitions (Rep)	23 (50%)	21 (33.3%)	13 (40.6%)	18 (25.7%)	75
2 Paraphrastic Reformulations (PR)	14 (30.5%)	21 (33.3%)	10 (31.2%)	25 (35.7%)	70
3 PR involving definitions (Def)	7 (15%)	10 (16%)	4 (12.5%)	20 (28.6%)	41
4 Reformulation + specification/ expansion (SpEx)	2 (4.5%)	10 (16%)	5 (15.6%)	7 (10%)	24
Total	46	62	32	70	210
Occurrences per 10,000 words	63	66	71	37	52
Coverage	13/15	11/15	14/15	15/15	53/60

Some type of reformulation occurs in all but two of the 3MT talks, with deliberate repetition being the preferred strategy (50% of occurrences). Its main purpose is for rhetorical effect, underlining important points often using ternary repetition, a well-known trope of classical rhetoric:

(5) *Is it because of* environmental effects like diet and health? *Is it because of* the effects of a particular person like their age, ethnicity or immune system? Or *is it because of* the effects of the particular type of virus that they've been infected with? (3MT)

Four of the 3MT talks also make use of an "echo" strategy, where deliberate repetition serves to reinforce the main message. In a talk on vehicle efficiency, for example, the speaker repeats in her closing words the key concepts of her research:

(6) Opening utterance: Today I'm going to talk to you about *improving vehicle efficiency using springs.*
Closing utterance: I'm helping to *improve vehicle efficiency*. That's all being done by *using some springs* (3MT)

The 3MT talks also use a number of PRs to simplify the audience's task of comprehension, but in line with the official 3MT instructions to contestants to 'avoid terminology,' reformulations involving more specialized terminology are sparse.

In the author videos, while all four categories are exploited, the coverage is uneven: four of the videos (very technical videos targeting a highly specialized audience) contain only one or no reformulation procedures, whereas others contain a large number, perhaps a further indication of the lack of genre stability already remarked upon in Section 3. In a video on the Lassa virus, for example, there are 10 occurrences of reformulation. The following extract from this video contains four PRs:

(7) S: But there's a few individuals that have *a very very diverse population of viruses within them*, **so there's lots of mutations and changes**. So this tells us that there are at least some people infected with Lassa who might have, who might be *chronic carriers*, **who have been carrying the virus for longer giving it more time to to mutate and change**. However most of these mutations that happen within a person because they help the virus at the immune system, then *are evolutionary dead ends*, euh **so where those mutations are not well suited to be transmitted to another person. So it seems like they occur but then you rarely see them again in another person.** (Video)

In (7), each italicized expression is subsequently rephrased (in bold) to make the key ideas clear, with the final sentence representing the second reformulation of *evolutionary dead ends*.

The author videos also contain repetitions but it is often difficult to draw the line between deliberate repetition and disfluencies. Several speakers repeat common evaluative adjectives and adverbs such as *very*, as in the first line of (7) above, or *incredible/incredibly* (ten occurrences in one short presentation). Although providing emphasis and illustrating the researcher's enthusiasm, in a way typical also of CMC discourse (Herring, Stein, and Virtanen 2013), these repetitions could also be prompted by the constraints of online speech processing and appear less rehearsed than the much slicker ternary repetition constructions and "echo" strategy noted in the 3MT talks.

The *SciAm* podcasts contain a high proportion of Rep (40.6%), some of which also involve ternary repetition for rhetorical effect:

(8) J: we're using it in every building, every bridge and every highway (*SciAm*)

Due to the high level of journalist involvement and the rather different genre expectations of popular science, repetitions are also used in other ways. Several of the podcasts contain plays on words, with a keyword connected to the topic being repeated to make a pun or joke. In a podcast on the discovery of Neanderthal skeletons, the journalist puns on the word "bones" at the end of the report: *No bones about it*. Likewise, in a podcast on windstorms the journalist picks up two keywords from the scientific report (rain and wind) to make a joke, alluding to two well-known Bob Dylan songs:

(9) J: And as this study shows: it won't just be that a hard rain's a-gonna fall.
We'll be blowing in the wind, too. (*SciAm*)

Lexis in these podcasts is very accessible, with practically no abstruse terminology, and therefore little need for definitions. There are however several examples of paraphrastic reformulations:

(10) S: this can be a problem for the plants, because they have a reduced ability to evolve resistance against diseases.
J: *Meaning that if we lose pollinators, it's the plants' genomes that may go to seed.* (*SciAm*)

In (10), the journalist reformulates the moderately technical explanation of the scientist (reduced ability to evolve resistance) into very everyday language, again using a pun with *seed*. Most such reformulative expressions are used by the journalists. The role of the researcher, as already mentioned, is in any case circumscribed in the *SciAm* podcast to just the occasional quote.

Reformulation strategies are proportionally less frequent in the *Nature* podcasts than in the other subsets. They contain however the highest proportion of reformulations involving definitions (28%). When combined with PRs, these two categories account for 65% of the occurrences (see Table 3), and are used by both journalists and scientists. In (11), it is the journalist who explains a term in advance so that listeners will be able to follow:

(11) J: Scientists refer to this tiny group as a reservoir (*Nature*)

Deliberate repetitions, however, are mainly used by scientists, to highlight important points:

(12) S: It is really one of the major questions that we have to answer in the future: *why, why expressing this* marker? *Why expressing this* signature? (*Nature*)

whereas repetitions revolving around plays on words for humorous intent are solely the preserve of the journalists, as in the *SciAm* subset. At the end of the podcast on Bird Beaks, for example, the two journalists engage in humorous banter with a series of puns involving birds:

(13) J1: I think she was winging it a bit, but I have no "egrets" about giving the story "top billing." If you want to find out more about that "egg-cellent" paper, then you'd better "flamin-go" and find it on the Nature website,
J2: And do feel free to Tweet us @NaturePodcast if you have any more "fowl" puns. (*Nature*)

4.2 Illustration procedures

Illustration overlaps to some extent with reformulation as both types of procedure enable speakers to communicate a specialized field of research to a less specialized audience in such a way that is relevant and understandable for them, preventing possible communicative failure. However, whereas reformulation involves some type of "retreatment" (Gülich 2003) or recycling, illustrations – comparisons and analogy, exemplification, metaphor, and scenarios – need not necessarily be employed to rework what has already been verbalized.

After briefly outlining their scope, we will compare their distribution across our corpus (Table 4).

1. Comparisons and analogy. Through comparisons with everyday objects and experiences, concepts can be made more accessible for the lay audience.
2. Exemplification. Exemplifications enable speakers to illustrate abstract or complex concepts with concrete examples, and to underline the applications of the research.

3. Metaphor: Adopting a traditional concept of metaphor as an implicit analogy of two unlike things, we identify those occurrences where speakers exploit the linguistic realizations of metaphor to reinforce the clarity or impact of their research.
4. Scenario. Following Ciapuscio, we use this this term to refer to the way speakers create "a possible yet imaginary situation that allows them to explain a complex event to the audience" (2003: 212).

Table 4. Illustration strategies

Types of strategy	3MTs	Author Videos	*SciAm* Podcasts	*Nature* Podcasts	*Total*
1 Comparison and analogy	9 (21.5%)	14 (48%)	8 (24%)	18 (43%)	49
2 Metaphors	4 (9.5%)	1 (3.5%)	8 (24%)	9 (21%)	22
3 Examples	21 (50%)	13 (44%)	17 (50%)	13 (31%)	64
4 Scenarios	8 (19%)	1 (3.5%)	1 (2%)	2 (5%)	12
Total	42	29	34	42	147
Per 10,000 words	57	31	75	22	36
Coverage	15/15	12/15	12/15	15/15	54/60

Examples are abundantly used in all four subsets and appear to be an essential strategy when addressing a non-specialized audience.

Illustration is widely used by the 3MT presenters ($n = 57$ per 10, 000 words), but it is the use of scenarios, a rarity in the other subsets, that appears to be a real feature of this genre. The student presenters make use of fairly elaborate scenarios involving imaginary situations in order to stage their talk. In a talk on natural language processing, for example, the speaker sets out a scene for the audience as if it were a scene from a film or play:

(14) Okay, so let me set the scene for you. You're driving in the desert. It's the middle of the night. There's no one but you on the road and then suddenly "bang" you see this thing crash landing in front of you. You decide to explore, so you take out your torch and you walk towards it and then you see these strange markings in the ground. (3MT)

The proportion of illustration strategies is lower overall in the author videos than in the 3MTs, with three of the very technical videos containing none. There are also practically no examples of scenarios or metaphors. There are however a number of comparisons and analogies with everyday objects to clarify matters (15):

(15) S: over time we realized that graphene behaves a lot *like paper*. (Video)

The *SciAm* podcasts contain proportionally the highest ratio of illustration strategies in our four subsets (75 per 10,000 words). Examples once again represent a high proportion as do comparisons and analogies (24%). In (16), the journalist and scientist work together using the same analogy of tying a shoelace in order to illustrate the problem of making molecular knots:

(16) S: "With a molecule you can't just grab hold of the ends and tie them like you would a shoelace. They're too small for that [...]."
J: [...] remember when you were learning to tie your shoes, your mom or dad put a finger in the middle of the knot, to make it easier to tie? Well, Leigh and his team did something similar, but used metal ions as the "fingers" to keep the knot tying organized. (*SciAm*)

Metaphors, on the other hand, are used exclusively by the journalists (and not the researchers): *a* ***cocktail*** *of common beehive viruses, concrete* ***recipes***.

Although the *Nature* podcasts contain fewer illustration strategies overall, comparison and analogy do feature and when combined with metaphors account for 64% of occurrences. Once again there is a divide in the way journalists and scientists use these strategies. As in the *SciAm* podcasts, metaphors (17) and the few scenarios used (18) are solely the preserve of the journalists:

(17) J: So hopefully one day researchers will carry out experiments like these, and the fairy circle mystery will finally *be put to bed*. (*SciAm*)

(18) J: *Imagine you're a virus particle*, floating outside your next bacterial host. You're preparing to infect the cell, injecting your DNA into the bacterium. (*SciAm*)

Scientists themselves appear to avoid these more literary or imaginative uses of language, preferring the more neutral strategies of comparison or exemplification. As seen above, this is not the case however in the 3MT presentations. Unlike the confirmed researchers, the PhD students make heavy use of dramatization (scenarios), which contributes to the creation of a "show" more than a research presentation.

4.3 Questions

Questions are engagement markers that presuppose the existence of an addressee and are inherently interactional, soliciting either direct interaction in the case of conversation and dialogue, or simulated interaction with the reader or listener in writing and spoken monologue. To investigate the degree of interactivity of the soundbites, we analyzed both the frequency and the functions of questions in the data. Our definition of questions included interrogative clauses and sentence fragments that concluded with a question mark in the transcripts or interrogative intonation in the recordings as in (i) and (ii), but excluded embedded or indirect questions:

i. What kind of patterns did you see?
ii. But how?

Table 5 shows that questions are used extremely often in the scholarly soundbites, averaging 43.5 per 10,000 words overall (vs only 0.9 per 10 k words in the written texts), by the scientists themselves in the monologues (all the 3MTs and all but two author videos) and mainly by the journalists in the podcasts. The subsets in which they are the most heavily used, both in frequency and in terms of coverage, are the 3MTs, which are delivered to a live audience, and the *Nature* podcasts, which are live interviews between scientist and journalist. In contrast, in the author videos and the *SciAm* podcasts, only about a third of the recordings contain questions, hence their lower frequency (around 20 per 10 k words).

Table 5. Number and frequency of questions in the corpus

	3MTs	Author videos	*SciAm* podcasts	*Nature* podcasts	Total
Number of Questions	35	17	10	113	175
per 10,000 words	48	18	22	60	43.5
used by scientists	35 (100%)	16 (94%)	3 (30%)	15 (12%)	69 (39%)
used by journalists	0	1 (6%)	7 (70%)	99 (88%)	107 (61%)
Coverage	11/15	5/15	6/15	15/15	37/60
Number in corresponding written texts*	0	3	9	7	19
per 10,000 words	0	0.4	1.5	1.1	0.9

(*) PhD Abstract in the case of the 3MTs, Research Article for the other 3 subsets

Our functional classification of questions broadly followed the distinction made by Thompson (1998) between audience-oriented and content-oriented questions but was adapted to our mix of monologue and dialogue data, with five categories:

1. Addressee-oriented questions
   a. Interactional: arousing interest in, or evoking a response from the interlocutor/audience.
   b. Clarification and checking: asking the addressee to repeat or clarify an utterance.

2. *Content-oriented questions*
   a. Research questions: raising scientific issues.
   b. Information-seeking/providing: asking the addressee to supply information (in dialogue) or introducing information by setting up question and answer pairs (in monologue).
   c. Confirmation-seeking: asking the addressee to confirm information (in dialogue).

The classification results are shown in Table 6.

Table 6. Functions of questions in scholarly soundbites

Function	3MTs	Author videos	*SciAm* podcasts	*Nature* podcasts
1. Addressee-oriented				
1a: Interactional	12	1	3	12
1b: Clarification and checking	0	2	0	5
2. Content-oriented				
2a: Research Questions	15	14	3	26
2b: Information-seeking	7	0	4	55
2c: Confirmation-seeking	1	0	0	15
Total	35	17	10	113

The 3MT speakers are characterized by their highly interactional stance: a third of their questions (1a: Interactional) are addressed directly to the audience, using the second person pronoun *you*:

(19) Would you believe me if I told you this was my brain on drugs? (3MT)

Practically all these uses occur in the opening and closing moves of the talk (Orientation and Termination, see Section 3), to arouse the audience's interest in the topic at the outset and to return to the live communicative context at the end. Content-oriented questions are also common, however, as speakers often present the research question of their thesis as a direct interrogative:

(20) Do some strains of HIV make people sick more quickly than others? (3MT)

They also use content questions to communicate information by setting up a pseudo-dialogue, in which they provide both the question and the answer, invariably introduced by the marker *Well*, thereby keeping tight control over the development of the discourse:

(21) So how do we find them? Well, we compute them. (3MT)

In contrast, questions are seldom used in the author videos, which adopt a primarily expository communicative style. There are no questions in 10 out of the 15 videos (including the two where a journalist intervenes), and only 3 are addressee-oriented. Research questions, all asked by the scientists, form the main category, but cluster heavily in a single video (containing 9 out of the 14) where the author's daring claims are contested by another researcher.

Despite the presence of two speakers, journalist and scientist, in all the *SciAm* podcasts, no real dialogue takes place here either. The speakers' respective roles are reflected in the questions asked: interactional questions are asked by the journalist to arouse audience interest, while research questions are asked by the scientists. Among the four information-seeking questions, two are addressed to the listeners, with the scientist's soundbite neatly dovetailed in to provide the answer:

(22) J: Eat like a caveman, no dairy, no grains, no sugar and so on. But what you probably won't find on many paleo plates today?
S: "Pine nuts and moss and tree bark and mushrooms."
J: Laura Weyrich, a paleomicrobiologist at the University of Adelaide. (*SciAm*)

The *Nature* podcasts show not only the highest frequency and coverage but also the widest range of question-types of our four subsets. Interactional questions, again all asked by the journalist, are addressed either to the listeners (*Ever heard of fossil water?*) or to a second journalist who acts as a foil in the opening banter (see Section 4.1):

(23) J1: Adam, what do you think is the toughest, most bad-ass organ?
J2: Maybe the liver (…)? Oh no, wait, maybe the brain (…)? Or the heart? (*Nature*)

The distinctive feature of this subset, however, is the very high proportion and number of information-seeking questions. The two interlocutors engage in constant dialogue, with the journalist using her questions to structure the interview, moving it forward step by step to cover the main points of the published study, and the scientist providing the information in her answers:

(24) J: What's so great about molluscs? (…) How did they all evolve? (…) Why was this particular fossil so mysterious? (…) What are we calling this new creature? (…) This guy [=fossil] just has a sort of little shell on its head. Is that useful for it? (…) And why is the fact that it has this spiky radula tongue so important? (…) (*Nature*)

When the journalist considers it necessary for listeners to understand, she asks the scientist to confirm her summary or formulation:

(25) J: Your team found really striking differences between Neanderthals, right?
S: Yeah absolutely. (*Nature*)

This is the only subset where this function is used to a significant extent. In these carefully structured interviews the journalist's questions and her requests for confirmation as the dialogue progresses enable the knowledge asymmetry to be overcome. This maieutic approach is very similar to that analyzed by Ciapuscio (2003) in his study of interviews between a science journalist and the researchers in preparation for the writing of a popular science article.

4.4 Scientists' comments on their work

An interesting feature of the soundbites compared to the written data is that in addition to presenting the research itself, speakers also make personal comments on their ongoing research. Basing our analysis on a close reading of the scripts we identified and classified the different expressions (see Table 7). In 3MTs, two types of comment are dominant: the challenge or difficulty of the research, probably because the doctoral students are still struggling to overcome some obstacles:

(26) Building the robot is really only a small part of the challenge. The difficulty lies in its control. (3MT)

and enthusiasm:

(27) Am I excited about my project? Yeah, I really am. (3MT)

No comments of this type feature in the PhD abstracts.

Table 7. Scientists' comments on their work.*

Types of comment	3MTs	Videos	*SciAm* Podcasts	*Nature* Podcasts
challenge, difficulty	8	6	1	5
enthusiasm, excitement, fascination, joy of discovery, delight, satisfaction, interest	8	19	1	21
surprise, puzzlement, incomprehension, uncertainty, disbelief, incredulity	0	14	0	7
Other (1 occ. each)	relief, sadness	0	0	confidence, competition, discomfort
Total	**18**	**39**	**2**	**36**
Coverage	14/15	11/15	2/15	14/15

(*) expressed as the number of occurrences

In the other three subsets, the difficulties of the research account for only a small proportion of comments, presumably because as the research is published, the main challenges have been overcome. In the author videos and *Nature* podcasts, positive comments predominate, no doubt for the same reason (successful completion of the work), ranging from expressions of excitement and the joy of discovery:

(28) S: We were all very excited. I had to go for a walk to calm myself down. So yeah it wasn't a Eureka moment but it was definitely you know a we did it moment. (Video)

to wonder at natural phenomena (described as "mind-boggling"), and even the fun of science:

(29) S: for us it's like playing Lego; it's magnetic Lego. And now we can play with this atomic Lego. (*Nature*)

These positive comments are in stark contrast to those we found in our study of the laboratory notebook, a daily record of the early stage of research, where negative comments expressing anger, frustration, perplexity, and self-criticism dominated (Rowley-Jolivet and Carter-Thomas 2016). The surprise element in scientific discovery also crops up quite frequently, reflecting the puzzle-solving nature of scientific activity:

(30) S: the thing that surprised us is that (…) We were really kind of shocked because we were thinking how does one single channel control two opposing things (*Nature*)

and scientific controversy is on occasion expressed very directly, without the conventional hedging and politeness strategies characteristic of the RA:

(31) S: My first reaction on reading the paper was No, this is wrong, something's wrong. (Video)

The exception is the *SciAm* podcasts. As the journalist does most of the talking and the scientists' comments are reduced to very brief quotes, this leaves very little room for them to express their attitudes to their work. We found only a single occurrence each of enthusiasm and difficulty.

5. Genre implications

As mentioned in the Introduction, the ubiquity of digital media in all spheres of society has led to their integration into the very functioning of some social institutions and to the adoption of media logics by these institutions, a process referred to as mediatization. Science has traditionally been a fairly autonomous, closed subsystem with its own codes of conduct and validation, requiring a high level of expertise to access the knowledge produced, unlike other subsystems oriented towards the general public such as politics, sport or education, where mass media are a useful channel of interaction with the target audiences. The transmission of scientific knowledge towards the general public was conventionally accomplished by science popularization, predicated on the deficit model in which science communication is conceived of merely as a vehicle to 'translate' science for a lay audience. The structural impact of the media on the scientific domain could therefore be expected to be lower than in other fields, and indeed, Rödder and Schäfer back in 2010 found that science's media resistance was quite high.

Things appear to be changing, however. The four sets of soundbites studied here are all produced within contexts that are highly institutionalized (higher education in the case of 3MTs) or strongly market-oriented (major publishing houses in the case of the podcasts and videos). These subsystems seem to be undergoing a process of mediatization, which may in turn impact on the genres produced by the scientists themselves. Universities, whether for economic and political reasons (to attract funding, to demonstrate their social utility…), or because they are constrained to do so by the global competition between universities triggered by the Shanghai and other ranking systems, or for more diffuse reasons such as their immersion in a media-rich society, now seek increased media coverage of their activities, setting up their own PR departments, and as shown by the 3MT™ initiative of the University of Queensland, engage in market-oriented practices such as branding and reputation management. As argued by Pallas and Wedlin (2013), ranking tools lead to measures of scientific value that are based more on media logics such as simplification (using citation metrics), standardization (the same ranking is applied worldwide), and popularization, than on traditional academic criteria of worth. Very similar media logics appear to be at work in the 3MT competition. The scientific content is drastically reduced and simplified, the presentations are highly standardized and the adjudicating criteria the same worldwide, and speakers overtly seek to catch the audience's attention and tell a good story. Such skills appear far removed from those currently valued in other university genres and it is debatable whether they would pull much weight with a thesis examining board, more inclined to scrutinize the candidate's mastery of theory and the robustness of the methodology and the results, than be told an engaging story. In this respect, 3MTs may potentially constitute negative training for the viva – or on the contrary, they may gradually have an impact on the way the viva genre itself is conducted in the future, infecting the jury's expectations and the candidate's presentation towards a more promotional style. The 3MT competition, by introducing and valuing these skills at a very early stage in the research career of scientists – during their doctoral research – may have a long-term impact on the mediatization of research.

Our other three subsets (podcasts and author videos) are all hosted or produced by major publishers, who are already engaged in media activities. The digital revolution has led them, however, to adapt and innovate, often radically, in the platforms and formats they offer in order to increase their visibility and promote their knowledge products. The huge expansion in the type and size of audience, with research that was previously confined to an esoteric circle of peers now available to the global audience, has clearly been taken on board. The podcasts, broadcast daily or weekly, can be automatically downloaded to the user's mobile device via an RSS feed "so you can listen when on the train, walking the dog or sitting in the garden" (*Nature* Podcast site), a 'push' strategy that brings the scientific findings directly to users worldwide. Posted on the popular YouTube channel, the author videos can likewise be viewed worldwide, as can the 3MT talks. In media logic, attention-getting is considered a form of social capital, and a channel such as YouTube enables the accumulation of such capital on the Internet.

As our chosen term of 'soundbites' indicates, both journals and researchers have also understood the need for brevity when communicating via the Internet, managing to condense dozens or even hundreds of pages of research into a few seconds or minutes in order to convey the gist of the work. They have also risen to the challenge of switching from an initially written mode of communication (theses, RAs) to the spoken mode. This switch involves not only a change in mode however, which researchers are accustomed to when an RA is presented as a conference talk or vice versa. In the case of the soundbites, it also involves managing without the traditional warrants used in research genres to ground and support claims, i.e. the literature review and citations, and the scientific visuals. In the absence of these warrants, considerable recontextualization is necessary, and speaker ethos, communicative skills and audience involvement can be expected to take on more importance.

This is indeed what we have observed in this study. As our move analysis has shown, certain traditionally important facets of research that would be difficult to grasp or uninteresting for the lay audience, such as the theory or methods, are simply omitted or glossed over. Striking adaptations are also found in the discursive and linguistic strategies of recontextualization used (scenarios, metaphor, dialogic interactivity through questions…) when compared to those of the corresponding research genres. Furthermore, a side to research that is erased from RAs and theses, but that these soundbites give the scientists freedom to express, is that of their personal comments during the research process. This feature may be related to the medium itself, as the prevalence of the expression of emotions in CMC has been observed in several online contexts (blogs, Twitter, YouTube comments, etc.; see e.g. Cislaru 2015), as has a trend towards personalization.

It is perhaps too soon to know what lasting implications these changes may have on the evolution of scientific genres, but some conjectures can be made. The picture of scientific research that the average listener would gather from these recordings may be a partial, or even biased, one. The topics chosen for web diffusion, whether video or audio broadcasts, appear to shun highly theoretical subjects, doubtless because they are considered inaccessible to the global audience. This may in the longer term induce an imbalance in the types of research made available to the digital audience compared to those that circulate in traditional genres. The radical reduction in the scientific content masks the complexity of research, in which many methods or setups must be tried before things can be made to work (Rowley-Jolivet and Carter-Thomas 2016). Moreover, all the soundbites have an important promotional purpose. The author videos and podcasts follow the publication of a research article, and serve to increase its visibility – and to promote the researchers themselves, who speak in person in the recordings. The very brevity of the data studied here reinforces this promotional intent. In another possible indication of the mediatization of science, the traditional dichotomy between information and entertainment appears to be collapsing in these soundbites: to attract and keep the attention of listeners, both sets of podcasts contain punning and plays on words, by the journalists at least, while 3MT speakers make many (often self-deprecating) humorous remarks to engage their audience, and could therefore be considered "infotainment." It will be interesting to see to what extent this media logic continues to gain force and affect science communication, and which of the four types of soundbite discussed will survive long-term and become an addition to the genre set available to researchers.

References

Bhatia, Vijay K. 2005. "Generic Patterns in Promotional Discourse." In *Persuasion across Genres: A Linguistic Approach*, ed. by Helena Halmari, and Tuija Virtanen, 213–225. Amsterdam: John Benjamins. https://doi.org/10.1075/pbns.130.13bha

Campagna, Sandra, Giuliana Garzone, Cornelia Ilie, and Elizabeth Rowley-Jolivet (eds). 2012. *Evolving Genres in Web-mediated Communication*. Bern: Peter Lang.

Carter-Thomas, Shirley, and Elizabeth Rowley-Jolivet. 2017. "Open Science Notebooks: New Insights, New Affordances." *Journal of Pragmatics* 116: 64–76. https://doi.org/10.1016/j.pragma.2016.12.003

Ciapuscio, Guiomar E. 2003. "Formulation and Reformulation Procedures in Verbal Interactions between Experts and (semi-)Laypersons." *Discourse Studies* 5: 207–233. https://doi.org/10.1177/14614456030050002004

Cislaru, Georgeta. 2015. "Emotions in Tweets: From Instantaneity to Preconstruction." *Social Science Information* 54(4): 455–469. https://doi.org/10.1177/0539018415597793

Garland, Ruth, Damian Tambini, and Nick Couldry. 2018. "Has Government been Mediatized? A UK Perspective." *Media, Culture & Society* 40(4): 496–513. https://doi.org/10.1177/0163443717713261

Gilbert, G. Nigel, and Michael Mulkay. 1984. *Opening Pandora's Box: A Sociological Analysis of Scientists' Discourse*. London: Cambridge University Press.

Gülich, Elisabeth. 2003. "Conversational Techniques Used in Transferring Knowledge between Medical Experts and Non-experts." *Discourse Studies* 5: 235–263. https://doi.org/10.1177/1461445603005002005

Hammarfelt, Björn, Sarah de Rijcke, and Alexander D. Rushforth. 2016. "Quantified Academic Selves: The Gamification of Science through Social Networking Services." *Information Research* 21(2): eSM1.

Herring, Susan, Dieter Stein, and Tuija Virtanen (eds). 2013. *Pragmatics of Computer-mediated Communication*. Berlin: Mouton de Gruyter. https://doi.org/10.1515/9783110214468

Hu, Guangwei, and Yanhua Liu. 2018. "Three Minute Thesis Presentations as an Academic Genre: A Cross-disciplinary Study of Genre Moves." *Journal of English for Academic Purposes* 35: 16–30.

Jacobi, Daniel. 1999. *La communication scientifique: Discours, figures, modèles*. Grenoble: Presses Universitaires de Grenoble.

Knorr-Cetina, Karin D. [1981] 2013. *The Manufacture of Knowledge: An Essay on the Constructivist and Contextual Nature of Science*. Amsterdam: Elsevier.

Latour, Bruno. 1987. *Science in Action*. Cambridge, Mass.: Harvard University Press.

Myers, Greg. 2003. "Discourse Studies of Scientific Popularization: Questioning the Boundaries." *Discourse Studies* 5: 265–279. https://doi.org/10.1177/1461445603005002006

Myers, Greg. 2010. *The Discourse of Blogs and Wikis*. London: Continuum.

Pallas, Josef, and Linda Wedlin. 2013. "Governance of Science in Mediatized Society: Media Rankings and the Translation of Global Governance Models for Universities." In *Global Themes and Local Variations in Organization and Management: Perspectives on Glocalization*, ed. by Gili S. Drori, Markus H. Llerer, and Peter Walgenbach, 295–308. New York: Routledge.

Pennec, Blandine. 2006. La reformulation en anglais contemporain: indices linguistiques et constructions discursives. PhD Dissertation, Université Rennes 2.

Pérez-Llantada, Carmen. 2013. "The Article of the Future: Strategies for Genre Stability and Change." *English for Specific Purposes* 32: 221–235. https://doi.org/10.1016/j.esp.2013.06.004

Pérez-Llantada, Carmen. 2016. "How Is the Digital Medium Shaping Research Genres? Some Cross-disciplinary Trends." *ESP Today* 4 (1): 22–42.

Rödder, Simone, and Mike S. Schäfer. 2010. "Repercussion and Resistance. An Empirical Study on the Interrelation between Science and Mass Media." *Communications* 35 (3): 249–267. https://doi.org/10.1515/comm.2010.014

Rowley-Jolivet, Elizabeth, and Shirley Carter-Thomas. 2016. "La vraie histoire de la recherche expérimentale? Comparaison entre la narration de la recherche dans les cahiers de laboratoire et dans les articles de recherche." In *La mise en récit dans les discours spécialisés*, ed. by Catherine Resche, 97–120. Bern: Peter Lang.

Swales, John M. 1990. *Genre Analysis. English in Academic and Research Settings*. Cambridge: Cambridge University Press.

Swales, John M. 2004. *Research Genres. Explorations and Applications*. Cambridge: Cambridge University Press. https://doi.org/10.1017/CBO9781139524827

Thompson, Susan. 1998. "Why Ask Questions in Monologue? Language Choice at Work in Scientific and Linguistic Talk." *Language at Work* 13: 137–150.

Ventola, Eija, Celia Shalom, and Susan Thompson (eds.). 2002. *The Language of Conferencing*. Bern: Peter Lang.

CHAPTER 6

Continuity and change

Negotiating relationships in traditional and online peer review genres

Ruth Breeze
University of Navarra, Spain

The move to new open access formats in scientific peer review is thought to have influenced the way referees and authors interact. This chapter considers the genre of author responses to referees in terms of relational work, and compares published responses in an anonymized online open access review system with a similar corpus of confidential responses to referees submitted by traditional means. In general, the online open access responses are longer, more syntactically complex and show more evidence of relational work than the confidential responses. However, when analyzed qualitatively, the proportions of different compliant and non-compliant moves are found to be similar in the two corpora. This chapter thus illustrates continuity and change in one genre's transition from traditional to open, online format.

Keywords: peer review, open access, author response, online genres, scientific writing

1. Introduction

Peer review has come to be regarded as an essential part of the process of scientific publication, and an important element in the ongoing development of science itself. However, it also has been criticized for lack of transparency and possible unfairness. In the quest for the optimum evaluation system, some scientific journals are now exploring different forms of open peer review that have become possible thanks to digital media. One such journal is *eLife*, a journal of biomedical and life sciences, which has opted to publish both the referees'[1] reports and the

1. In this chapter, for the sake of clarity, I use the term "referee" to mean the person who has written the peer review report/comments, and "author" to refer to the person who has written the response. I have avoided the term "reviewer" on the grounds of similarity with "peer review."

https://doi.org/10.1075/pbns.308.06bre
© 2019 John Benjamins Publishing Company

authors' responses in the Internet alongside the finished article itself. This makes this journal's website a privileged window onto the peer review process as such. At the same time, it opens up new possibilities for researching possible changes in the way that referees and authors interact when they know that their comments and responses are going to be made available to a wider audience online. In other words, we can observe how the genres of peer review evolve in response to the "opening up" through the affordances of Internet of what was previously a private exchange. Study of this situation sheds further light on how genres evolve across the digital transition.

After explaining the move toward different forms of open access in scientific peer review, this chapter will focus specifically on the genre of responses to referees in terms of relational work. With a view to tracking developments in this genre across the digital divide, I present a qualitative and quantitative analysis of two corpora of responses to peer review in biomedical sciences, one from a journal that supports anonymized open peer review published on the journal website alongside the finished article (*eLife*), and the other consisting of confidential responses to peer review obtained locally.

2. Ongoing developments in peer review

Although scientists have always commented on and reacted to each other's work, the formal institution of peer review came into being when scientific journals were established during the eighteenth century (Spier 2002). Peer review in one form or another is now regarded as a standard part of the publication process worldwide, and enjoys an almost mythical status (MacDonald 2014): having a good peer review system is regarded as the ultimate guarantee of legitimacy, as a "quasi-sacred process that helps to make science our most objective truth teller" (Horton 2000: 148). However, this does not mean that it is entirely unproblematic, and even staunch defenders of the peer review system admit that there are many practical issues involved. The processes involved are intrinsically slow and time consuming. The difficulty of locating referees and persuading them to put in several hours of work without recompense is a perennial problem in the current highly competitive research environment. Moreover, the specialized nature of many scientific research fields means that peer review is rarely completely "blind." Issues such as bias, nepotism or vested interests have cast shadows over the fairness of the review process (Wold and Wennerås 1997; Groves 2010; Lee et al. 2013), unfair treatment of non-native English writers is still an issue (Pérez-Llantada 2012: 132; Zare et al. 2016: 173), and the claim that peer review discourages innovation and fosters mediocrity casts a shadow in some circles: "what is published in top journals is what there is least reason to reject" (MacDonald 2014: 271).

To address some of these points, journals have recently started to question their established practices in this sense (Godlee 2002). One of the innovations intended simultaneously to give credit to referees for their work and lessen the scope for injustice is the introduction of open peer review, which has been made possible by the affordances of the Internet, particularly the opportunity this opens up for adding extra material to the journal website (Van Rooyen et al. 1999; Ford 2013, 2015). In the last ten years many journals have started to include the peer review comments and author feedback for published articles on their websites, with a view to making the review process more transparent.

However, before going further, it is important to note that the blanket term "open peer review" itself gives rise to a certain degree of confusion. In contrast to established concepts like single-blind and double-blind peer review, open or "transparent" peer review means different things to different journals. In some publications, it involves simply providing the referees' names with the final published version of the paper, while in others it involves publishing both referee names and also their full reports, as well as the authors' responses to those reports (Rodgers 2017). In others, it implies different degrees of transparency, leading some authors to claim that there are 22 distinct forms of "open peer review" (Ross-Hellauer 2017). These range from the suspension of anonymity (by revealing either the authors' names to the referees, or both authors' and referees' names to each other, at the initial review stage), to large-scale involvement of the wider scientific community (through systems such as platforms to share first drafts within the wider expert community for voluntary peer review, or so-called proof-reading queues in which volunteers can read papers in exchange for providing constructive corrections). Open peer review now also sometimes includes the option of open, direct two-way reviewer-author or reviewer-reviewer discussion (Ross-Hellauer 2017). In most cases, however, the new approach to peer review is either completely "open and non-anonymous" in the sense of revealing the names of the referees (i.e. *British Medical Journal*, see Groves 2010), or it is "open" in the sense of publishing the referees' reports in full or edited form online, without giving referees' names unless they specifically consent, as in the case of *eLife* (Shekman, Watt, and Weigel 2013), the example used here. Before setting out the design of the present study, I will present a brief overview of the ongoing changes in peer review as it crosses the digital divide, first sketching in the way these issues are framed within the scientific community, and then describing the way they have been approached from an applied linguistics perspective.

2.1 Professional perspectives on open and confidential peer review

The opening up of the peer review process in these ways has been widely discussed in the bibliography of the disciplines involved (Rodgers 2017; Ross-Hellauer 2017). Greater openness in the peer review process, in one form or another, seems in general to be welcomed by the scientific community, and particularly by the younger generation (Ross-Hellauer, Deppe, and Schmidt 2017). Several supposed advantages have been proposed, although issues of confidentiality and scientific credibility mean that it is actually rather difficult to prove the extent to which any of these are really bearing fruit. The main advantage of publishing the referees' reports online would seem to be that this greater transparency should force referees to focus on the merits of the paper itself, and therefore encourage them to be fairer, more objective, and to justify their decisions better. This would discourage referees from allowing factors such as gender, author position or nationality to influence the decision to publish. Adopting an optimistic view, Groves (2010) believes that, on the whole, the move towards publication of reviews makes them likely to be of higher quality, which means that they will probably be more useful to the authors. As DeCoursey (2006) says, publishing reports should "encourage reviewers to do a thorough job." Along similar lines, in a more open system referees will not be able to "block" papers as a result of prejudice, personal dislike or competing interests – or at least, they will need to provide compelling reasons for doing so. On the negative side, open peer review presents one obvious difficulty, which is that when the referees are named in the case of an unfavourable review, there is a risk that the people whose work has been criticized will react badly and perhaps even take revenge at a later date (DeCoursey 2006; Khan 2010). The upshot of this may be that referees may be less willing to be critical (Rodgers 2017), a view that has received some support from a recent survey in which 50% of respondents were found to believe that open identities would make peer review worse or much worse (Ross-Hellauer, Deppe, and Schmidt 2017). Khan (2010: 6425) asserts that such reports will be of lower quality: "open review can cause reviewers to blunt their opinions for fear of causing offence and so produce poorer reviews." For related reasons, non-anonymity may also make it harder to recruit referees (Rodgers 2017). Although in general, greater transparency should mitigate rather than exacerbate the problems of rivalries or prejudice by putting the review process on a more professional footing, the solution is not perfect. By placing greater pressure on reviewers, open peer review might occasionally even slow down the review process and thereby cause extra difficulties for editors. In short, the main focus of research and discussion from the disciplines themselves has been on differences between open and conventional peer review, on author and reviewer preferences, and on the payoff between openness and confidentiality in terms of scientific rigour.

2.2 Genre perspectives on open and confidential peer review

Looking at the issue from an applied linguistics perspective, it is clear that the types of texts that occur regularly in the specific context of peer review constitute genres, in the sense of a set of "choices which represent effective ways of getting things done in familiar contexts" (Hyland 2003: 22). They contain the "clusters of register, style, lexis, and other features" that "reflect the different personal and institutional purposes of writers, the different assumptions they make about their audiences, and the different kinds of interactions they create with their readers" (Hyland 2003: 23). Of course, genres are also constrained by the medium in which they are produced, and it has been amply demonstrated that when genres make the transition from print to digital format they usually undergo certain changes (see Luzón and Pérez-Llantada this volume). On the basis of the foregoing discussions within the scientific community, we might assume that when comments and responses are produced publicly online, rather than in complete confidentiality, certain changes are likely to take place in both. From the general trends in thought outlined above we can assume that the transition from private to digital, public and open is likely to mean that both comments and responses will be more elaborate and more clearly argued. They may well also pay greater attention to interpersonal aspects.

The professional discussions within the scientific community on open access peer review rarely focus on aspects of linguistic realisation, although they do occasionally touch on aspects of this. For example, Khan (2010) surmises that reviewers may take greater care with the way they express their judgments if they know that their comments are going to reach a wide audience, and may tone down their criticisms. Respondents to the survey by Ross-Hellauer, Deppe, and Schmidt (2017: 23) thought that published reports (whether signed or not) would be likely to offer less strong criticism and would show less "disinhibition" on the part of referees, although it is not clear whether they regarded this as positive or negative. Along similar lines, but taking a positive angle, DeCoursey (2006) thinks that "reviewers would be more tactful and constructive." However, none of these writers offers evidence that this is actually the case. Moreover, there is no mention here as to how the authors' responses might change in response to this different style of report.

Within applied linguistics there has so far been relatively little interest in peer review genres, probably because of the general difficulty of obtaining texts for research purposes. What there is generally focuses on the referees' comments, rather than on the responses. In 1996 Swales defined the referee report as an "occluded genre," because at that time this was considered to be a private document, intended only for the immediate users (editors, referees and authors). Kourilová

(1996) was able to examine commands, doubt, suspicion and criticism in medical peer review reports, and found that the style in her sample was generally direct and unmitigated. Gosden (2001, 2003) examined the structure of reports, and provided a list of categories found in referees' comments (see Table 1). Fortanet-Gómez (2008) confirmed the general structure identified by Gosden (2003), and developed a formal taxonomy for analyzing referee reports based on a three-way distinction between what she terms criticisms, recommendations and questions. Using Hyland's (2005) framework for stance, Fortanet-Gómez and Ruiz-Garrido (2010) provided a comparative analysis of boosters, hedges and attitude markers in business organization and applied linguistics reports, finding that attitude markers were used with a similar frequency, but that both boosters and hedges were more common in applied linguistics reports, which seemed to be both longer and more carefully negotiated with regard to reducing the face threat to authors. Finally, Mungra and Webber (2010) examined referee comments on a small sample of texts submitted by Italian authors to medical and scientific journals, and developed a taxonomy of comments divided into two broad categories: content and language. They found that 56.28% of the comments were related to content, and that these fell into categories 1–10 in Table 1, while the rest were about language, discourse and organization (category 11).

Table 1. Classification of referees' comments in Gosden (2001) and Mungra and Webber (2010)

A. Asking for further explanation or clarification	1. Incomplete literature
	2. Lack of association between claim and data
B. Asking for technical details to be provided	3. Procedural infelicities and lack of rigor
C. Asking for strength of claim to be modified	4. Explain why data are unusual
	5. Scientific reasoning errors of own data
D. Asking for new references, or asking for references to be corrected	6. Terminology or definitions
	7. Statistical irregularities
E. Questioning the appropriacy of the article for the journal itself, or the section of the journal	8. Incorrect scientific interpretation of other authors
	9. Lack of association between claim and prior research
	10. Sampling errors
	11. Language use and discursive/rhetorical aspects of the text
Gosden (2001)	Mungra and Webber (2010)

Concerning the authors' responses, these would seem to be even more "occluded" than the reports themselves, and have accordingly attracted even less research interest. In fact, the only source of information about these is to be found in various comments within articles about peer review reports that have implications for the authors concerned. Kourilová (1996) stressed the interpersonal dimension of comments and responses, proposing that the Slovak writers of scientific English should learn pragmatic skills and socio-cultural strategies to cope better with this daunting task. Mungra and Webber (2010) also surmised that Italian authors sometimes found referee comments intimidating or confusing, and suggested that they should be encouraged to try to uncover the intention behind the comment itself. Gosden (2001) emphasizes the power imbalance between referees and authors, and stresses that if authors wish to get into print (which they do), they will acquiesce to the referees' wishes with little, if any, resistance. Gosden (2001) is unique in applying the Swalesian concept of "move," that is, a structural segment that has a specific interpersonal communicative function and purpose. He draws up the following list of moves found within the genre of authors' responses to the referees' comments:

- Providing a technical detail.
- Discussing – clarifying the information provided previously.
- Downgrading claims made previously.
- Adding and/or correcting references.
- Responding to challenges by making arguments in favour of publication.
- Stating that language changes have been made.

However, evidence from peer review practices in other areas of research suggests that this list is incomplete. For example, Belcher (2007) emphasizes the importance of displaying willingness to continue revising and resubmitting in the face of extensive critical commentary. Yet she also observes that authors – particularly those who are not familiar with the culture of peer review – need to learn that requests for change may be negotiable, since this should color the way they respond to certain requests. Given the lack of a robust framework, the present study will proceed from a qualitative discourse analysis of authors' responses, in which categories of moves are identified on the basis of reiterative reading and comparison of the text. It will then support this with quantitative study of salient features, following principles of corpus-assisted discourse analysis.

3. Theoretical framework

To gain a wider perspective on the issue of peer review and responses, it is useful to step outside the immediate framework of EAP (English for Academic Purposes)-oriented research and look at the interpersonal phenomenon of peer review from the perspective of pragmatics. For the authors, submitting their work to the peer review process obviously involves voluntary exposure to a face threat, as well as offering the possibility of genuine support to improve their article. The face threat involved is both general (criticism of the article as a whole) and specific (criticism can be aimed at the authors' method, results, analysis, statistics, etc., their assessment of previous authors' contributions, and even their understanding of science). To address this question in my analysis of the authors' responses, I will draw on the concept of relational work developed by Locher and Watts (2008). In their understanding, relational work refers to the interpersonal level of communication, describing the "work" that individuals invest in negotiating relationships with others (Locher and Watts 2008: 78). They note that relational work "comprises the entire continuum of verbal behaviour from direct, impolite, rude or aggressive interaction through to polite interaction" (Locher and Watts 2005: 11). Developing this further (Locher and Watts 2008: 96), they state that relational work is "all aspects of the work invested by individuals in the construction, maintenance, reproduction and transformation of interpersonal relationships." It is thus a broader construct than "face" or "facework," which is mainly concerned with mitigating FTAs (although here that is undeniably important). Relational work is an essential part of our social behaviour that often goes unnoticed. It is part of what we are used to and what we expect to occur. Moreover, when we communicate in writing, such relational work is clearly performed through the vehicle of genres, i.e. text types characterized by particular organizational and linguistic features in which professional/institutional and private purposes are pursued in a socially recognized, conventionalized way.

The present chapter thus seeks to address the following research question: how does the transition from confidential to open peer review change the genre of the response to referees? This is approached first from a qualitative perspective, by identifying and characterizing the different moves used and determining their prevalence. These findings are triangulated by a quantitative analysis of salient features indicating the relational work undertaken in the text.

4. Corpus and methodology

With a view to conducting a comparison between author responses in traditional confidential peer review and those found in new online open peer review formats, it was necessary to obtain two corpora of peer review reports and author responses.

In the case of open peer review, I chose to work with *eLife*, a high-profile open access journal covering biomedical and life sciences. The *eLife* reviewing system is characterized by controlled openness: reviewers and editor discuss their comments with each other openly in order to reach a consensus, and then the editor compiles all the requirements into one "lightly edited" decision letter (Shekman, Watt, and

Weigel 2013, *eLife* 2016). Both the consensus decision letter and the author's reply are ultimately published online, as is the documentation from any further stages in the revision process, alongside the article itself. Although the referees have the option of revealing their identity, most choose to remain anonymous. For the present study, I assembled a sample of 30 referee reports and responses to reviewers from the 2018 editions of *eLife*, taking ten each from *Biochemistry and Chemical Biology*, *Cell Biology* and *Epidemiology and Global Health* (*eLife* corpus).

To represent traditional non-open peer reports and responses, I obtained 30 referee reports and responses to reviewers generated locally, from Spanish authors' submissions to international non-open-peer-review journals in biomedical sciences (NOPR corpus). Informed consent was obtained from the scientists who donated the reports and responses (we should note that it is not possible to obtain informed consent retrospectively from anonymous peer reviewers), and the documents were all fully anonymized before use. Descriptive data about these two corpora can be found in Table 3.

First, the reports and responses were read. Although initially I had intended to apply Gosden's categories (see above) to analyze the authors' responses, the substantial mismatch between these and the responses in my corpora meant that it was necessary to devise a more appropriate coding system. For this reason, a process of reiterative qualitative reading was used to identify the main moves in these responses (see Airey 2011), and to code for these. The proportions of different moves in each corpus were calculated.

At this stage, I also identified a rather disparate set of features that appeared to be salient in relational or textual terms. These included:

- Relational features: interpersonal aspects such as reference to self and referees; the functions of thanking and apologizing; verbs for framing author statements and actions; stance features such as hedging devices for mitigation, and adverbs for boosting; verbs and adverbials for marking changes made.
- Textual features: length, organization and complexity of response, materialized in discourse markers and part of speech analysis.

It should be noted that these interpersonal features are strongly associated with relational work in the text, whereas textual aspects concerning length, organization and complexity of the text are linked more indirectly to this, but are still important (as discussed above, the fact that the responses are published encourages authors to negotiate the extent of their compliance more carefully, and this also has relational aspects). The corpora were uploaded to SketchEngine for quantitative analysis, and searches were conducted to test the frequency of the features identified as salient above, in the spirit of corpus assisted discourse analysis. To avoid repetition, the rationale for including each of these features will be discussed below, alongside the account of their frequency, and the explanation of their role in the interpersonal or textual dimension of the text.

5. Results and discussion: Qualitative move analysis

Given the nature of the response genre, it is clear that most relational work focuses on compliance or non-compliance with the referees' report. Although previous work suggested that most of the responses would indicate compliance in various forms (adding details, lessening the strength of claims, clarifying text, making language corrections, etc.) (Gosden 2001), examination of the present corpora revealed that these were far from being the aspects of the responses that offered most interest. Compliance was often expressed very concisely, using formulae such as: "as requested by reviewers," or "we have added the information suggested." However, the category of "negotiating non-compliance" proved much richer in terms of relational work, yielding four sub-categories. Finally, the category "clarification" was also interesting, because it often seemed to be used as a face-saving device by authors, suggesting that the supposed "mistake" detected by the referees was, in fact, a misunderstanding. In Table 2, the three main categories and their subcategories found in both corpora are illustrated with examples:

Table 2. Classification of moves in author responses, with corpus examples

Move type 1: Compliance. Comply with referees' comments by making corrections or adding further information

- This was a key omission in the original submission and we appreciate the opportunity to correct this.
- We have now expanded upon our description.
- We have toned down the results.

Move type 2: Negotiate non-compliance
2a. *Express inability to make the changes required.*

- Given the essential cell cycle functions of both phosphatases, siRNA experiments unfortunately are impossible to conduct.
- Unfortunately, technical limitations do not allow the protein stoichiometry experiment that the reviewer suggests.
- We did not examine liver/body ratios in pre-surgery mice, because we do not have enough mice.

Table 2. (*continued*)

2b. *Refuse to comply but offer alternative/compromise solution*

- Rather than removing the data completely, we propose to move this 'endogenous' experiment to the supplementary data (….). We hope that this is a suitable solution.
- The referees suggest an alternative method of looking for hysteresis, but it is not obvious for us that it would enhance our ability to detect it. We very much hope to pursue this question in the future studies once we upgrade our optical trap system.
- In respect of the inclusion of p-values, we understand the debate and are aware of their potential misuse in science. However, we believe there is value to the inclusion of both confidence intervals and p-values in the manuscript. In response to this comment we have now systematically added confidence intervals to all our effect size estimates reported in the paper as well as in the tables and have retained the p-values.

2c. *Refuse to comply, providing reasons to support this refusal*

- While we agree with both reviewer #1 and #2 that precision is normally a more useful measure of the power of a trial design, in the face of such fundamental imprecision I strongly believe it is not meaningful here.
- We agree that incomplete labelling is an important consideration for accurate measurements of distance. However, we don't believe that unreacted cysteines are a likely source of incomplete labelling for the following reasons: (…..). We now state this more clearly in the manuscript.
- With due respect to the referee, we believe that the title of the manuscript reflects the most remarkable result of our study and in this sense, we would like to maintain it.

2d. *Comply while indicating disagreement*

- We still consider this to be the most likely scenario and feel it is important to raise this point, but we are happy to tone down our statements. In the Abstract, we altered this sentence to say that "RNA shredding […] may thus be a reflection…"

Move type 3: Allege misunderstanding on the part of referees

- It seems that the way we presented our data was confusing because major conclusions were not understood. Consequently, we introduced major changes in the organisation of the Results section.
- On reflection, we realize that the distinction between how, and why, models have been used to inform field and experimental designs was not made clearly or powerfully enough in the original manuscript.
- We understand that our arguments were not explicit enough and consequently we have made substantial modifications to the manuscript.

Graph 1 illustrates the number of times each of these moves was found in the *eLife* and the NOPR corpora.

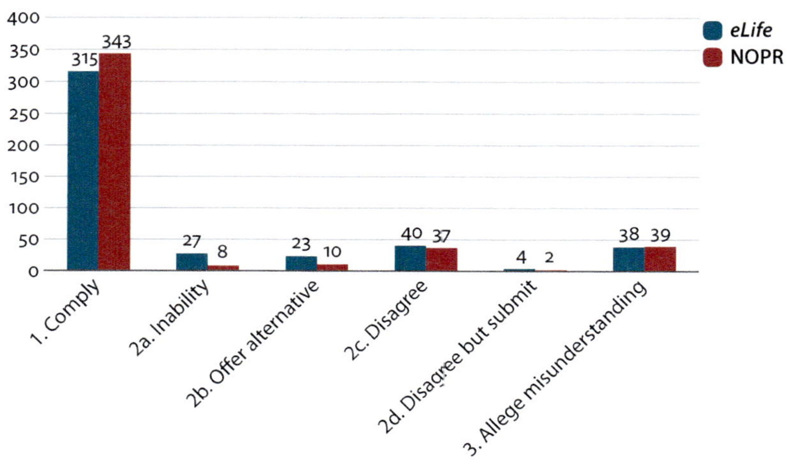

Graph 1. Number of moves in author responses in eLife and NOPR corpus

The total number of moves was similar in the two corpora (447 in *eLife* and 439 in NOPR). As Graph 1 illustrates, there was a slightly larger proportion of compliance (move 1) in NOPR. However, the amount of direct disagreement (2c) was effectively the same in both corpora. Where the authors in *eLife* differed was in their greater use of moves 2a (inability to make the changes required) and 2b (offering an alternative to complete compliance). Only the differences for 1 (comply) and 2a (inability) were found to be significant using the Chi square test ($p < 0.01$).

The following longer example from *eLife* shows how incompliance is negotiated in a context of alleged inability (category 2a):

> We shared the reviewers' wish to interrogate the phosphorylation and ubiquitylation status of the NOLC1 mutants (….) However, such experiments are cost-prohibitive (…) Our previous work had shown that 293T cells can be used to study stem-specific E3 ligases and their regulation (….) Thus we hope that the reviewers agree that the insight gained by repeating ubiquitylation studies in hESCs does not justify the large financial investment.

We may note that the authors launch their response by mitigating what they are about to say, indicating their personal desire to comply. They then explain in impersonal terms the reasons why they did not. This is followed by a justification based on their own previous work (which the referees are tactfully assumed not to have known about before, and which provides an answer to the problem that underlies the call for more experiments). In relational terms, the authors are not defying the referees in a new face threat, but rather providing the referees with a good opportunity to accept this act of non-compliance and save face at the same time. The final sentence is phrased as a polite request ("thus we hope that the reviewers agree") with a repeated emphasis on "the large financial investment."

The relational work in responses of this kind arguably allows the refusal to comply with the referees' request to be accepted by the referees without loss of credibility or authority, and is particularly characteristic of *eLife*. Arguably, this points to one way in which the author response genre is evolving in response to the digital transition: authors are now aware that they are explaining themselves to a wider potential audience, and are willing to invest more time in defending their own position and negotiating non-compliance with demands made in what they feel to be unfair or ill-informed reports.

6. Results and discussion: Quantitative analysis of salient features

As explained above, quantitative analysis was used to examine how the type of relational work illustrated in the previous section pervaded the response corpus as a whole. As Table 3 shows, both the referee reports and the responses were longer in *eLife* than in the NOPR corpora, which confirms the hypothesis that the knowledge that one's report is destined for publication is likely to encourage both referees and authors to take a greater effort to justify their points.

Table 3. Descriptive data from *eLife* and NOPR corpora: Referee reports and responses

	eLife	**Non-open-peer-review (NOPR)**
No. of tokens / referee reports	35,130	30,786
Mean no. of tokens per referee report	1,171	1,026
Mean sentence length / referee reports	29.7	23.3
No. of tokens / responses	49,597	40,532
Mean no. of tokens per response	1,653	1,351
Mean sentence length / responses	29.5	25.9

The analysis that follows is based on a quantitative exploration of the features that emerged from the qualitative analysis as having salience in interpersonal or textual terms. In each case, the features are discussed in the light of relevant bibliography, and related to the overarching theme of relational work.

6.1 Interpersonal features

With a view to gaining a better overview of the relational features in the two corpora of responses, the following interpersonal features identified as salient during the qualitative analysis were examined using corpus tools: reference to self and other participants (i.e. referees) (Hyland 2005); thanking and apologizing

(Locher and Watts 2008); verbs used to represent authors' actions and opinions (Thompson and Ye 1991; Hyland 2000; Locher and Watts 2008); hedging (modals and verbs expressing degrees of caution) and boosting (adverbs) (Hyland 2005); attitude markers such as (potentially evaluative) adjectives (Hyland 2005); and change markers.

The presence of participants (self and others) in the text is a key marker of interpersonality which has been widely discussed in the literature. In his model of stance and engagement, Hyland (2005) includes self-mentions ("we") as aspects of stance, but separates them from reader mentions ("you"), which he classifies as an aspect of engagement. In the present case, I include both self-mentions and references to the referees together, in order to shed more light on the relational dimension of these texts.

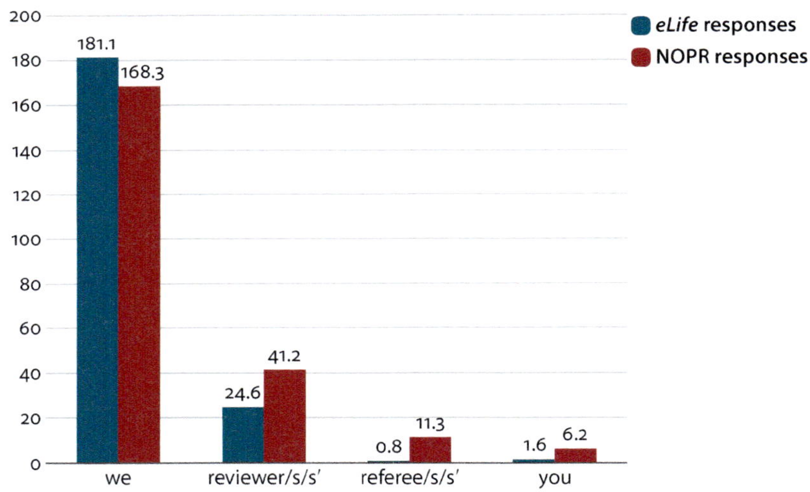

Graph 2. References to self and referees in *eLife* responses and NOPR responses (normalized to f/10,000 words)

As Graph 2 illustrates, references to the two parties chiefly involved in these highly relational texts were evenly balanced between the two corpora. In both cases the authors used "we" to indicate the actions they were taking to meet the referees' requests, and there was a general tendency to talk of the referees in the third person ("as the reviewers suggested"), echoing the way the referees refer to the authors in the third person in their reports. This highly conventionalized form of reference, with avoidance of "you," seems to be another of the features of the genre which enables both parties to relate to each other in a rather impersonal way and thereby save face. From the perspective of politeness and relational work, we should note that frequent phrases like "as suggested by the reviewer" convey a degree of mitigation, the underlying construction of the relationship being that the reviewer is "helping" the reviewed, and that the comments are merely "suggestions" that the authors are free to follow.

One of the most openly relational aspects of the responses to referees is the way that the authors address the referees openly by thanking them, or by apologizing (Graph 3).

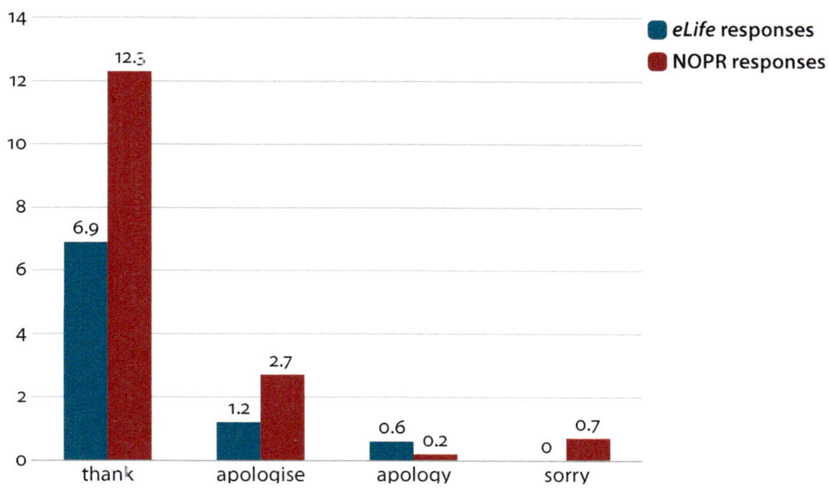

Graph 3. Thanking and apologizing in *eLife* and NOPR responses (normalized to f/10,000 words)

As in the texts analyzed by Johnson (1992) in a very different peer review context, the formulae used for thanking are highly conventionalized, and are found particularly at the beginning of the response. Here, the response genre (both public and private) often opens with a conventional phrase such as "We thank the reviewers for their careful review of our manuscript." In addition, conventional formulae such as "thank you for raising this point" or "we would like to thank the reviewer for pointing this out" are generally used to start a compliant response move, or to mitigate at the beginning of a non-compliant move. Other possibilities ("grateful," "gratitude," etc.) were absent. It is interesting that the NOPR authors thank the referees more, but it is possible that some of the repetitive instances of thanking, at least, were removed from the *eLife* authors' responses in the "light editing" undertaken to avoid repetition.

One feature that emerged as important here was the range of different verbs used with the subject "we" to present the acts undertaken by the authors. Graph 4 shows the normalized frequency of the commonest verbs of speaking, thinking, agreeing and doing found in the first personal plural in these texts, identified using frequency lists in SketchEngine.

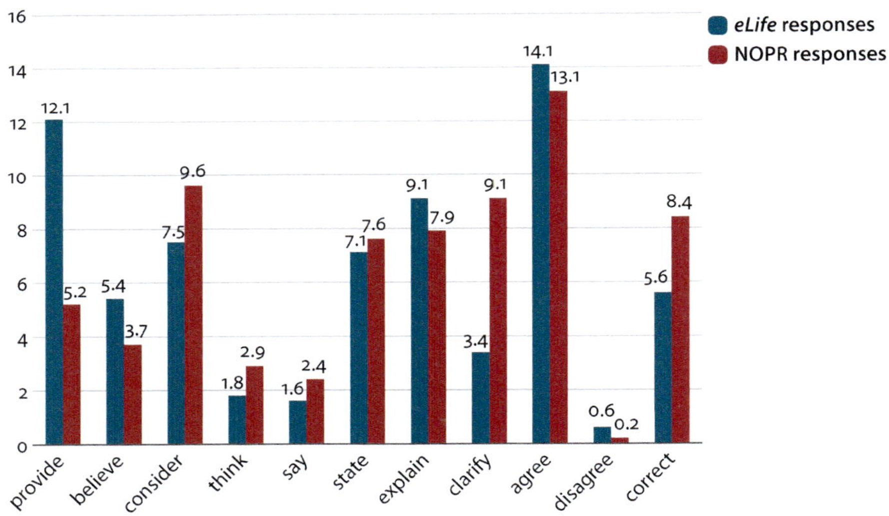

Graph 4. Verbs used to formulate author actions and opinions in *eLife* and NOPR responses (normalized to f/10,000 words)

These implicitly relational verbs reflect the way that the writer envisages his/her verbal interaction with the reader (Thompson and Ye 1991; Hyland 2000). The differences between the corpora are overall not statistically significant on Student's t-test. However, there is a small but interesting difference in the NOPR preference for "to correct," included here as a frequent verb describing the action undertaken by the authors at the referees' request, and which implies acknowledgement (in the confidential responses) that a mistake had been made. This contrasts with the *eLife* preference for "to provide," with its positive implication that the authors are now offering something extra that they had not previously included (rather than admitting that they failed to include something).

Hedging is one of the most widely studied features of academic discourse, because it is used for negative face redress, and as Hyland famously analyzed (2005), it is an essential element in the strategies that authors use to relate to their peers in academic publications (see also Fortanet-Gómez and Ruiz-Garrido 2010). Graph 5 illustrates the most frequent (verbal and adverbial) hedging devices found in these two corpora.

We may assume that hedging has an important role in these responses, where the level of confidence expressed by the authors has to be very carefully modulated in order to maintain (or restore) credibility, show respect for others, and manage the relationship with the referees (and in the case of *eLife* with the potential online readers). We should note that overall frequency of these hedging devices was greater in *eLife* (71.7 per 10,000 words as opposed to 43.1 per 10,000 words). This difference was found to be statistically significant ($p < 0.05$) using the paired

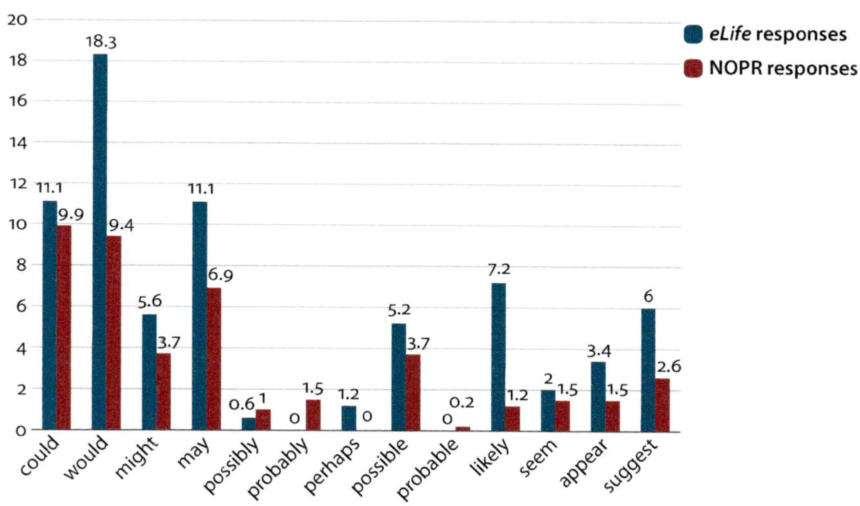

Graph 5. Hedging devices in *eLife* and NOPR responses (normalized to f/10,000 words)

Student's t-test. The explanation for this probably lies in the need for more careful relational work when the text is perceived to be open to public scrutiny.

Boosting is also widely acknowledged to play an important role in the interpersonal dimension of the text (Hyland 2005). Qualitative reading suggested that intensifying adverbs were by far the most frequent boosting device used in these responses. Graph 6 displays the most frequent types encountered in the two corpora.

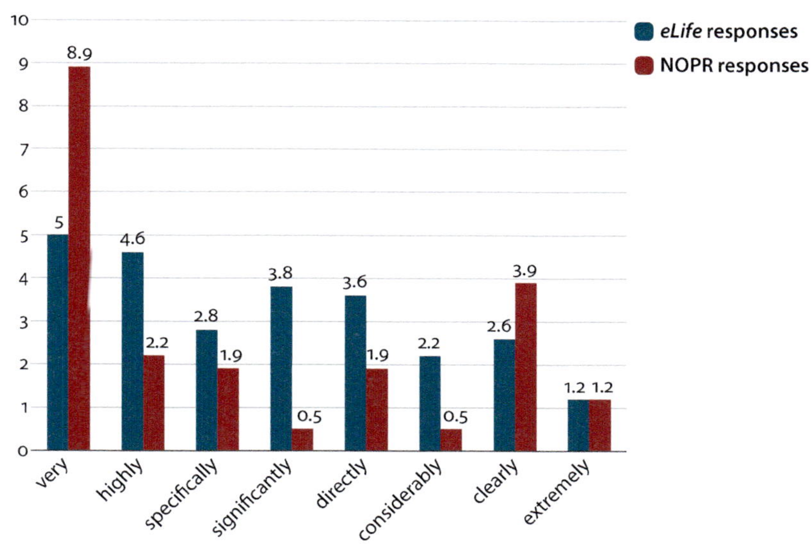

Graph 6. Intensifying adverbs in *eLife* responses and NOPR responses (normalized to f/10,000 words)

Graph 6 shows that with the exception of "very," which was nearly twice as frequent in NOPR, intensifying adverbs were much more frequent in *eLife* responses, a difference which was found to be just significant using the paired Student's t-test ($p < 0.05$). The open responses not only contained more intensification as such, but also had boosting of a more varied nature.

Finally, from the perspective of interaction, I looked at markers used to point out or emphasize changes made in the second version of the text (Graph 7). Interesting here is the discovery that "now" is much more common in the *eLife* responses corpus than in the NOPR responses corpus (27.6 compared to 13.8 per 10,000 words).

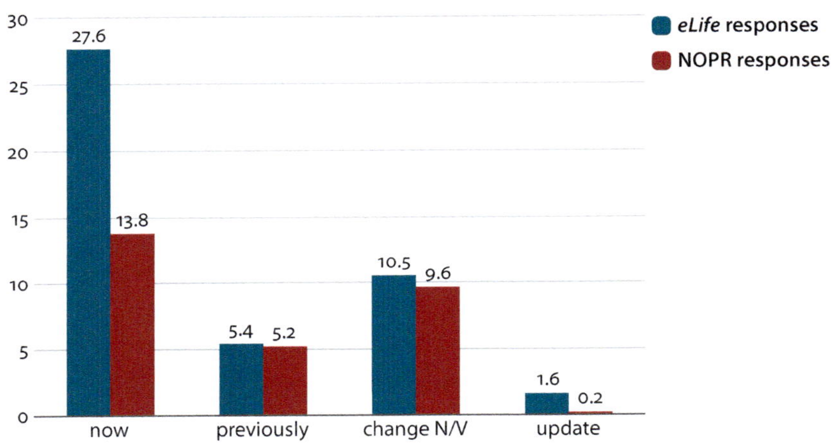

Graph 7. Possible markers indicating changes to revised versions in *eLife* and NOPR responses (normalized to f/10,000 words)

This seems to point to a more emphatic approach to reporting changes in the online format, using the adverb "now" to stress where alterations have been made. If we contrast these data with the verbs in Graph 3, above, we can observe that the NOPR authors often shape their responses in terms of "corrections," while the *eLife* authors prefer to emphasize the positive aspect of their modifications, using "now" to indicate what is new, in phrases such as "this is now better explained in the accompanying text" or "this is now stated in the methods." This might suggest a greater relational need to "save face" publicly in the online open responses, as compared to a stronger desire to show compliance in the confidential NOPR corpus.

6.2 Textual features: Length, organization and complexity

One of the hypotheses voiced by the scientists commenting on the move to open online peer review was that the reports would be more detailed, which would logically imply that the responses would also be longer and more complex. The word length data (see Table 3 above) suggest that this may indeed be the case. However, longer texts are not necessarily more complex or more sophisticated in terms of argumentation. In order to examine evidence for greater elaboration in the open peer review responses, I looked at discourse organizers (Hyland 2005). The results are shown in Graph 8.

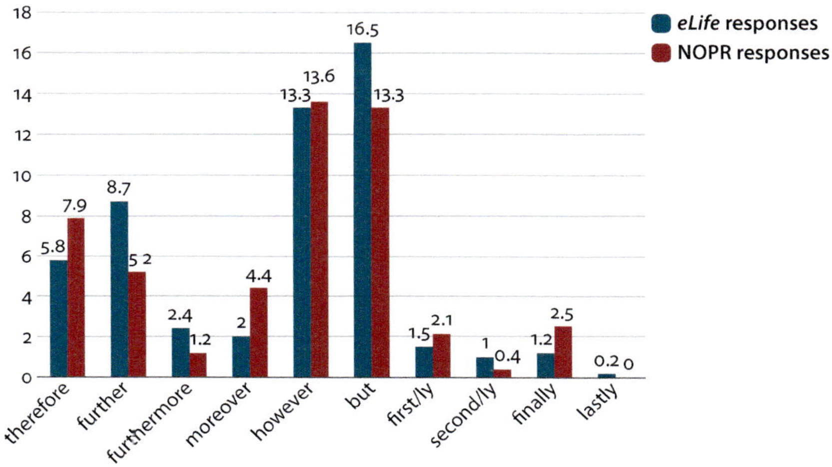

Graph 8. Textual organization devices in *eLife* and NOPR responses (normalized to f/10,000 words)

The markers of textual organization were found to occur with similar frequencies in both of the response corpora. However, since the general data from Table 3 also showed that sentences were longer in the *eLife* corpus, I decided to investigate sentence complexity, which can be taken as a further indication that these authors produced more elaborate responses. To this end, a part of speech (POS) analysis was conducted using SketchEngine, some key results of which are shown in Table 4.

In broad terms, the POS analysis uncovered very few relevant differences between the corpora, but it did show that the *eLife* responses contained 1.7 times as many "that"-subordinators, 1.7 times as many "wh"-determiners ("which" and "that") and 1.6 times as many when/where/how clauses as the NOPR responses, relative to corpus size (Table 4 shows both raw and normalized frequencies, and a keyness score indicating the ratio of occurrence in the two corpora). These results go hand in hand with the longer mean sentence length in the *eLife* responses, suggesting that responses intended for publication are likely to consist of more elaborate sentences containing more clauses, which is perhaps indicative of a more complex discursive style, greater conceptual elaboration and greater care with relational aspects.

Table 4. Frequency and keyness values of POS relating to syntactic complexity in *eLife* using NOPR as reference corpus

	eLife responses		NOPR responses		
Tag	Frequency	f/10,000	Frequency	f/10,000	Keyness score (ratio *eLife*:NOPR)
Wh-determiner	242	48.8	115	28.4	1.7
That subordinator	466	93.9	228	56.3	1.7
When/where/how	118	23.8	61	15.1	1.6

7. Conclusions

We have seen that the transition from confidential peer review to open online peer review has led to subtle but persistent changes in the author response genre, pushing it towards greater elaboration and a higher degree of attention to relational work.

In general, the genre of the response to referees involves complex relational work that operates on different levels. Responding to peer review, authors generally show compliance with the referees' wishes: they thank the referees, or apologize for mistakes, they correct the data/analysis, they supply missing information and/or they rectify their summary/discussion – as Gosden (2001) says, they seem generally to put up little resistance. Within this display of compliance, however, they may often extend their answers, in the desire to save face by showing that they fully understand the points made by the referees and have more to contribute. On the other hand, authors do have the option of taking the risky option of sticking to their guns and showing defiance. They can explain why they are unable to do what they have been asked, or they can bring in arguments showing that they are right (and the referees are wrong). Finally, they can also try to steer a middle course. In this, either they can maintain that they are in agreement with the referee, and that the apparent discrepancy is due to a misunderstanding, or they can state openly that they are not convinced by the referees' point, but that they will offer some kind of compromise, or (rarely) simply agree to differ but comply out of respect to the referee.

All of the above possibilities are available in both the private and the online response genre, and the authors make strategic use of them, using a range of relational and textual strategies to maintain the relationship with the referees while maintaining their own image of professionalism. It is striking that open peer review did not discourage these authors from defying the referees – in fact, disagreement moves were more frequent in the open online corpus, which gives some support to the view expressed by scientists that open peer review is empowering for authors

(Ross-Hellauer, Deppe, and Schmidt 2017). From a psychological point of view, we can speculate that there may actually be a greater incentive to defend oneself against a face threat made in the public arena.

Regarding other changes from confidential to open online peer review, the most noticeable quantitative difference identified here is the greater length and complexity of the open responses, which seem to indicate that openness is driving the genre towards greater elaboration. Advocates of open peer review in the scientific community felt that it would encourage higher quality reviews (DeCoursey 2006; Groves 2010), and the present evidence suggests that this has a knock-on effect on authors, encouraging them to produce more complex responses. We have seen that the responses to online open peer review are longer and more elaborate, and pay greater attention to the relationship that is being constructed with the referees. In terms of this relational work, the most striking finding was the increase in hedging in the open response corpus, which adds support to the idea that relational aspects are gaining more prominence here, and authors are devoting more effort to managing their relationship with referees, and with the putative external readers. We can surmise that this is influenced by the fact that the power balance, previously weighted heavily in favor of the referees, has subtly shifted. There is now an external "jury" of readers who can judge for themselves whether the referees' requests are justified. The authors still need to acknowledge courteously the referees' points, as before, but now they also have to ground their own responses more carefully, so that it does not seem as if they are simply yielding to pressure in order to get published. The higher degree of relational work required is manifested in greater discursive complexity and a higher degree of hedging and boosting. Other evidence of differences in relationship maintenance, such as the higher frequency of thanking and apologizing in the confidential responses, is less conclusive.

The limitations of the present study are clear. The corpora are small, and the NOPR corpus, though collected in a principled way, may not be entirely representative of pre-open access peer review. Nonetheless, it provides useful pointers for future research, as well as offering information that could be used to devise possible pedagogical applications.

References

Airey, John. 2011. "Talking about Teaching in English. Swedish University Lecturers' Experiences of Changing their Teaching Language." *Ibérica. Journal of the European Association of Languages for Specific Purposes* 22: 35–54.

Belcher, Diane. 2007. "Seeking Acceptance in an English-only Research World." *Journal of Second Language Writing* 16 (1): 1–22. https://doi.org/10.1016/j.jslw.2006.12.001

DeCoursey, Tom. 2006. "Perspective: The Pros and Cons of Peer Review." *Peer to Peer*. Last accessed June 14, 2019. http://blogs.nature.com/peer-to-peer/2006/06/perspective_the_pros_and_cons.html

eLife. 2016. "Increasing Transparency in *eLife*'s review process." *eLife*. September 2016.

Ford, Emily. 2013. "Defining and Characterizing Open Peer Review: a Review of the Literature." *Journal of Scholarly Publishing* 44 (4): 311–326. https://doi.org/10.3138/jsp.44-4-001

Ford, Emily. 2015. "Open Peer Review at Four STEM Journals: An Observational Overview." *F1000Research* 4:6. https://doi.org/10.12688/f1000research.6005.2

Fortanet-Gómez, Inmaculada. 2008. "Evaluative Language in Peer Review Referee Reports." *Journal of English for Academic Purposes* 7 (1): 27–37. https://doi.org/10.1016/j.jeap.2008.02.004

Fortanet-Gómez, Inmaculada, and Miguel Ruiz-Garrido. 2010. "Interacting with the Research Article Author: Metadiscourse in Referee Reports." In *Constructing Interpersonality. Multiple Perspectives on Written Academic Genres*, ed. by Rosa Lorés-Sanz, Pilar Mur-Dueñas, and Enrique Lafuente-Millán, 243–254. Newcastle: Cambridge Scholars.

Godlee, Fiona. 2002. "Making Reviewers Visible: Openness, Accountability and Credit." *Journal of the American Medical Association* 287 (21): 2762–2765. https://doi.org/10.1001/jama.287.21.2762

Gosden, Hugh. 2001. "'Thank You for Your Critical Comments and Helpful Suggestions': Compliance and Conflict in Authors' Replies to Referees' Comments in Peer Reviews of Scientific Research Papers." *Ibérica. Journal of the European Association of Languages for Specific Purposes* 3: 3–17.

Gosden, Hugh. 2003. "'Why not Give the Full Story?': Functions of Referees' Comments in Peer Reviews of Scientific Research Papers." *Journal of English for Academic Purposes* 2: 87–101. https://doi.org/10.1016/S1475-1585(02)00037-1

Groves, Trish. 2010. "Is Open Peer Review the Fairest System? Yes." *British Medical Journal* 341. https://doi.org/10.1136/bmj.c6424

Horton, Richard. 2000. "Genetically Modified Food: Consternation, Confusion and Crack-up." *Medical Journal of Australia* 172 (4): 148–9.

Hyland, Ken. 2000. *Disciplinary Discourses: Social Interaction in Academic Writing*. London: Longman/Pearson.

Hyland, Ken. 2003. "Genre Based Pedagogies: A Social Response to Process." *Journal of Second Language Writing* 12: 17–29. https://doi.org/10.1016/S1060-3743(02)00124-8

Hyland, Ken. 2005. "Stance and Engagement: A Model of Interaction in Academic Discourse." *Discourse Studies* 7: 173–192. https://doi.org/10.1177/1461445605050365

Johnson, Donna. 1992. "Compliments and Politeness in Peer Review Texts." *Applied Linguistics* 13 (1): 51–71. https://doi.org/10.1093/applin/13.1.51

Khan, Karim. 2010. "Is Open Peer Review the Fairest System? No." *British Medical Journal* 341: 6425. https://doi.org/10.1136/bmj.c6425

Kourilová, Magda. 1996. "Interactive Functions of Language in Peer Reviews of Medical Papers Written by Non-Native Users of English." *UNESCO ALSED LSP Newsletter* 19 (41): 4–21.

Lee, Carole J., Cassidy R. Sugimoto, Guo Zhang, and Blaise Cronin. 2013. "Bias in Peer Review." *Journal of the American Society for Information Science and Technology* 64 (1): 2–17. https://doi.org/10.1002/asi.22784

Locher, Miriam, and Richard Watts. 2005. "Politeness Theory and Relational Work." *Journal of Politeness Research* 1 (1): 9–33. https://doi.org/10.1515/jplr.2005.1.1.9

Locher, Miriam, and Richard Watts. 2008. "Relational Work and Impoliteness: Negotiating Norms of Linguistic Behaviour." In *Impoliteness in Language*, ed. by Derek Bousfield, and Miriam Locher, 77–99. Berlin: Mouton de Gruyter. https://doi.org/10.1515/9783110208344

Macdonald, Stuart. 2014. "Emperor's New Clothes: The Reinvention of Peer Review as Myth." *Journal of Management Inquiry* 24 (3): 264–279. https://doi.org/10.1177/1056492614554773

Mungra, Philippa, and Pauline Webber. 2010. "Peer Review Process in Medical Research Publications: Language and Content Comments." *English for Specific Purposes* 29: 43–53. https://doi.org/10.1016/j.esp.2009.07.002

Pérez-Llantada, Carmen. 2012. *Scientific Discourse and the Rhetoric of Globalization. The Impact of Culture and Language*. London/New York: Continuum.

Rodgers, Peter. 2017. "Peer Review: Decisions, Decisions." *eLife*. September 2017. https://doi.org/10.7554/eLife.32011

Ross-Hellauer, Tony. 2017. "What is Open Peer Review? A Systematic Review." *F1000 Research* 6: 588. https://doi.org/10.12688/f1000research.11369.2

Ross-Hellauer, Tony, Arvid Deppe, and Birgit Schmidt. 2017. "Survey on Open Peer Review: Attitudes and Experience among Editors, Authors and Reviewers." *Plos ONE*, December 2017: 1–28. https://doi.org/10.1371/journal.pone.0189311

Shekman, Randy, Fiona Watt, and Detlef Weigel. 2013. "Scientific Publishing. The *eLife* Approach to Peer Review." *eLife* 13 (2). doi: https://doi.org/10.7554/eLife.00799

Spier, Ray. 2002. "The History of the Peer-Review Process." *Trends in Biotechnology* 20 (8): 357–358. https://doi.org/10.1016/S0167-7799(02)01985-6

Swales, John M. 1996. "Occluded Genres in the Academy: The Case of the Submission Letter." In *Academic Writing: Intercultural and Textual Issues*, ed. by Eija Ventola, and Anna Mauranen, 45–58. Amsterdam: John Benjamins. https://doi.org/10.1075/pbns.41.06swa

Thompson, Geoffrey, and Yiyun Ye. 1991. "Evaluation in the Reporting Verbs used in Academic Papers." *Applied Linguistics* 12: 365–382. https://doi.org/10.1093/applin/12.4.365

Van Rooyen, Susan, Fiona Godlee, Stephen Evans, Nick Black, and Richard Smith. 1999. "Effect of Open Peer Review on Quality of Reviews and on Reviewers' Recommendations: A Randomised Trial." *British Medical Journal* 318: 23–7. https://doi.org/10.1136/bmj.318.7175.23

Wold, Agnes, and Christine Wennerås. 1997. "Nepotism and Sexism in Peer Review." *Nature* 387 (6631): 341–343. https://doi.org/10.1038/387341a0

Zare, Javad, Vahid Mahmoudi-Gahrouei, Saeed Ketabi, and Zahra Keivanloo-Shahrestanaki. 2016. "English for Research Publication Purposes: The Case of Scholarly Peer Review Comments." *Ibérica. Journal of the European Association of Languages for Specific Purposes* 32: 153–177.

CHAPTER 7

The multimodal bridge between academics and practitioners in the *Harvard Business Review*'s digital context

A multi-levelled qualitative analysis of knowledge construction

Carmen Daniela Maier and Jan Engberg
Aarhus University

In this chapter we study adjusting strategies employed in the transition from research-based articles to other generic forms when the same knowledge is constructed and (re)mediated for different audiences and purposes. The empirical evidence is represented by a series of articles and related digital materials about and for the practice of management from the digital context of the *Harvard Business Review*.

We adopt a multimodal approach in our qualitative analysis, addressing the way semiotic modes interact and are combined across genres to identify the roles of the semiotic interplay in the adjusting strategies employed for enabling the knowledge construction and (re)mediation processes. Furthermore, we demonstrate how the level of complexity of the knowledge communicated in the articles and their digital context may be investigated.

Keywords: multimodal academic knowledge, academic discourse, digital material, research knowledge construction and mediation, explanatory depth levels

1. Introduction

In our ongoing project, to which this contribution belongs (see, for instance, Maier and Engberg 2013, 2014; Engberg and Maier 2015, 2019), we analyze multimodal knowledge communication texts mainly from the core academic field, defined as "the communication of domain-specific knowledge generated by researchers for other researchers in order for them to understand and use that knowledge"

https://doi.org/10.1075/pbns.308.07mai
© 2019 John Benjamins Publishing Company

(Maier and Engberg 2016: 225). This approach is connected to Hyland's definition of academic discourse as aimed at "educating students, demonstrating learning, disseminating ideas and constructing knowledge," but also at "construct[ing] the social roles and relationships which create academics and students and which sustain the universities, the disciplines, and the creation of knowledge itself" (Hyland 2009: 1). In today's university world, especially the function of disseminating ideas outside the core field of academic discourse gains importance, making it relevant to widen the scope of academic knowledge communication from the core of researchers' communication with researchers to include also other audiences than researchers themselves. An example of texts from within the widened scope are the digital texts at the website of *Harvard Business Review* (https://hbr.org). Harvard Business Publishing, the organization behind *Harvard Business Review*, linked to Harvard Business School, "serves as a bridge between academia and enterprises" and Harvard Business Review "aims to provide professionals around the world with rigorous insights and best practices to help lead themselves and their organizations more effectively and to make a positive impact" (https://hbr.org/corporate/about).

Texts from this website will be the object of the present study. Hence, we will analyze a type of knowledge communication in the academic field where both senders and receivers are experts with an academic background, who however do not pursue the core aims of academic discourse of Hyland. Our study thus presupposes the relevance of the idea of a stratification of specialized discourse with different textual conventions, different linguistic and multimodal choices in the different strata – stratification of specialized communication (Hoffmann 1985; Kalverkämper 1990), and diastratic variation (Picton and Dury 2017). In both cases, the basic idea is that choices of linguistic means in specialized communication are different according to the level of expertise of the parties involved. The approach of Picton and Dury (2017) further distinguishes between parties with the same level of education, but different purposes with the expertise (e.g. academics and practitioners with academic degrees in the field of management). The purpose of our study is to investigate patterns in this stratum of academic discourse in the multimodal interplay between elements of the website and establish potential differences between articles and their surrounding digital material (transmedial gradation) under the assumption that articles and this surrounding digital material have different audiences and/or fulfil different purposes.

2. Conceptual frameworks

In this section, we will introduce the two basic conceptual frameworks that we draw upon in this chapter: knowledge mediation processes and levels of explanatory depth. Both frameworks aim at describing multimodal texts as a basis for mental knowledge construction processes and especially how the combination of modes is productive in this context. The first approach (knowledge mediation processes) is procedural and focuses upon the interaction and the relations between different modes in the process of creating the textual meaning. The second approach (levels of explanatory depth) is more structural and aims at assessing the complexity of the knowledge base created by the multimodal text.

2.1 Knowledge mediation processes

We problematize in this study the concept of knowledge communication by explaining how the contemporary digital context characterized by flexibility and instability has called for nuanced discussions of new knowledge mediation processes that take place when knowledge is constructed and disseminated.

For our present purposes, this problematization implies a discussion of two main knowledge mediation processes: the *knowledge expansion* and the *knowledge enhancement* processes (Engberg and Maier 2015; Maier and Engberg 2016). The first category of knowledge mediation processes concerns the breadth of the knowledge that is mediated while the second category of processes concerns the depth of knowledge. The categories might be considered as overlapping when not taking into consideration what differentiates them, namely this double perspective that one should take when exploring such mediation phenomena. We need to illuminate both their magnitude by revealing the knowledge expansion processes and their profundity by revealing the knowledge enhancement processes. Certain excerpts of multimodal data can be classified as belonging to sub-categories of both main processes, but not all of them. Thus, this is where the relevance of the present conceptualization resides as it enables us to reveal and argue for the necessity of this double perspective when exploring HBR complex communication strategies. In the following paragraphs, we will define and clarify the conceptual framework. In the subsequent analysis examples of the processes and knowledge types are found.

The *knowledge expansion processes* are related to specific strategies of combining *core knowledge* with two types of *peripheral knowledge*. The core knowledge is the contribution made to the central topics examined during the research work (according, for example, to an article's title and abstract). The peripheral knowledge includes both *background knowledge* and *supplementary information*. Through mediating background knowledge, the researchers' necessary and presupposed background expertise is revealed. The supplementary information includes items relevant for further topic-specific reading and research.

The *knowledge enhancement processes* ensure that new knowledge is facilitated by various dissemination formats. These processes can be both *evident* and *apparent*. Evident knowledge enhancement processes can comprise, for example, hyperlinks to the authors' other research works, or hyperlinks to related products (i.e. reports and books). The apparent ones can be manifested through facilitating access to the same knowledge in other dissemination formats (for example, the research article in a PDF format), or to unrelated knowledge from advertising materials embedded in the same digital context. For example, the hyperlinked imperative placed in a box on the right side of the "Neuroscience of thrust: Setting leaders up for success" webinar's screen: "With email newsletters get the latest HBR business insights directly to your box. Sign up today" (HBR 2017). This excerpt of multimodal data is categorized as providing an apparent knowledge enhancement process due to the fact that it does not contribute to an actual enhancement of the knowledge communicated in the webinar.

2.2 Levels of explanatory depth

Where the first analytical framework investigates processes of knowledge creation across modes and is procedural in its approach, the second framework to be presented here is more structural, aimed at assessing the level of complexity of the textual basis for knowledge construction. It is based upon the concept of explanatory depth developed in psychology. The point of departure is an observed *Illusion of Explanatory Depth* (Rozenblit and Keil 2002). It reflects a tendency of people in communicative interactions to overestimate the degree of detail of their explanatory knowledge on specific concepts. The work cited above has shown experimentally that university students tend to denote their explanatory understanding of the functioning of, e.g. a helicopter as good, although they are actually not able to subsequently describe the actual causes leading to the ability of a helicopter to hover and lift off vertically. Actually, non-expert explanatory knowledge structures are more skeletal and fragmentary than similar expert structures. But as they fulfil the requirements of explanation in non-expert situations, holders are functionally satisfied with their knowledge. Hence, when asked whether they can explain how a helicopter works, the answer will often be yes. And only when put into a situation where a higher level of detail is contextually required (e.g. through more detailed questions in an experimental setting) does the holder of the knowledge realize that the degree of explanatory depth is not sufficiently high.

We will not go deeper into the psychological mechanism in this paper. However, the study of the Illusion of Explanatory Depth has led researchers to suggest a scale of different degrees of explanatory depth, which we find useful for our descriptive purposes. They distinguish between three types of skeletal explanatory knowledge structure: *Causal relevance*, *Causal powers* and *Causal relations* (Keil 2003: 675–680; Keil 2011: 254–255). Below, a short presentation of each type:

- Causal relevance: This explanatory relation is the coarsest of the three. "Coding of causal relevance does not encode specific patterns of causal interactions but rather a sense of what properties matter most in a particular domain" (Keil 2003: 675). An example is that shape is typically seen as mattering when explaining and categorizing hand tools. Hence, in such an explanatory structure, the holder knows that two characteristics of a concept are related to each other, matter functionally to each other. But no specification is made concerning the more specific nature of the functional relation.
- Causal powers: In this explanatory relation the holder knows not only that there is some kind of functional relation between two characteristics but knows also what kind of role one characteristic has upon the other. "I know that magnets have the ability to exert an attractive force on various metals but may know little about magnetism and the reasons that some metals make good magnets while others do not. We can think of this level as the first level at which distinct causal roles are attributed to properties" (Keil 2003: 678). The attributed roles are typically few and coarse (e.g. *prevent* or *support*), and there is no representation of a coherent system or mechanisms leading to the functional relations.
- Causal relations: In this type of relation, the knowledge structure has system character, but the system consists only of the main units and their functional relations. No insights concerning the internal functioning of these entities are present. "For many people, their mental representations of the causal relations for cars may largely be confined to knowing that they convey people from place to place on roads, that they are propelled by an engine whose output is increased by pressing on an accelerator, and that they are slowed down by brakes" (Keil 2003: 679).

All three types of explanatory structures are coarse and do not show the full complexity of expert explanations. However, they are satisfactory to users, due to their functionality.

In the case of our analysis, we are not measuring the depth of the explanatory structures of people. Instead, we assess the level of explanatory depth expressed in the text analyzed and thus the kind of knowledge structure that the user is supposed to construct based on the text. We cannot, and do not want to say anything about what knowledge structures the readers will actually construct based on the text, as such structure building is dependent upon the structures existing in the mind of the user when receiving the text. However, we can propose a substantiated hypothesis about the minimal level of explanatory depth and thus of the complexity of the knowledge structure to be constructed by the user in order to comply with the textual intentions of the author.

Following this review of conceptual frameworks, we introduce below our data selection methods and our framing methodology.

3. Data selection method and framing methodology

As our purpose is to conduct a fine-grained qualitative analysis, we have considered that a limited number of digital materials is sufficient for achieving the present research purposes. The choice of data has been primarily motivated by our intention to highlight the transmedial gradation in HBR's knowledge mediation and remediation[1] (Bolter and Grusin 2002) across several media. This transmedial gradation, namely the strategy of spreading the same research results across several media for similar and additionally different purposes, is manifested through specific digital genre clusters, namely through groups of different genres focused on the same topic, that are repeatedly employed on the HBR digital platform. In each of the chosen digital genre forms – three research articles, a video, an interview and a webinar – the presence of the author and of a certain degree of audience participation are changing, while multilayers of relations are established among these genres through various means of knowledge communication.

The selection of these digital genre forms has been performed in two stages after careful reading and rereading of digital materials appeared in HBR from 2015 to 2017. In the first stage of our data collection work, we have identified the main clusters of generic forms that reappear in the company of all research articles. In the second stage, we have selected those research articles and their accompanying genres that could give us the possibility to investigate in detail a series of recurrent patterns of transmedial gradation present in HBR's knowledge mediation and remediation across several media.

We have thus chosen three digital HBR articles and for each article the accompanying other genre: *Emotions and the Art of Negotiations* (2015) and its related interview, *Controlling your Emotions during a Negotiation* (00:05:54), *Getting to Sí, Ja, Oui, Hai, and Da. How to Negotiate across Cultures* (2015) and its related video, *Getting to Yes across Cultures*, (00:02:17), and *The Neuroscience of Trust* (2017) and its related webinar, *Neuroscience of Trust Setting Leaders up for Success*, (01:00:10). In the articles, the author's presence is manifested through the written words and a small close-up photo. The audience participation is manifested through the

1. Remediation refers to "the formal logic by which new media refashion prior media forms" (Bolter and Grusin 2002: 273).

written comments of the readers. In the genres that supplement the articles, the presence of the author is manifested differently. In the interview, the author is talking in front of the viewer's eyes and the audience does not explicitly participate, but the author's advising statements, and the repeated usage of the "you" pronoun indirectly acknowledge its presence. In the video, the author is present only in the references made by the voice-over presenter who explains aurally and through written words and animated graphics the research work done by the author. In the webinar, the author's presence is manifested through his voice-over explanations that accompany static composite images. The audience is indirectly participating through the questions posed to the author and mediated by the HBR's representative. Hence, already this first overview demonstrates differences in the presence of author as well as audience in the different elements. In the actual analysis, we will look deeper into these differences underlying the idea of transmedial gradation.

Apart from that, it should be mentioned that the analyses include all multimodal elements found on the webpages of the respective materials because each of the above-mentioned genres is embedded in the digital platform in such a way that the transmedial phenomena cannot be explored otherwise.

Acknowledging that "multimodality is growing fast" (Norris and Maier 2014) and in order to attain the level of analytical delicacy necessary for this research endeavor, data has been explored from a multimodal perspective (Kress 2010; Kress and van Leeuwen 2001; Iedema 2003). Consequently, the meaning-making processes have been explored at the level of several semiotic modes (written text, speech, still and moving images, and layout), at the level of their interplay and during the resemiotization of knowledge from one mode to the other (Iedema 2003). In this way, it has been possible to identify and interpret their distinct or overlapping roles and their particular relations with the communicative purposes of the respective digital materials in order to underpin the above mentioned transmedial gradation.

In our analysis in this chapter, we have combined the multimodal perspective upon knowledge construction processes leading to knowledge expansion and knowledge enhancement with the perspective of the explanatory quality and complexity of the knowledge to be constructed on the basis of the multimodal texts. Whereas we have carried out the analysis of knowledge expansion and enhancement processes on all of the investigated texts, the analysis from the perspective of explanatory quality and complexity is only carried out on one of the genre clusters (article + interview) due to the high level of analytical delicacy and to space constraints.

4. Findings

In what follows, the main findings of our analysis related to knowledge mediation processes in *Harvard Business Review*'s articles and their accompanying materials are presented (see Table 1). We have divided the analysis of the two types of material, as the articles are closest to actual academic research articles in their form (although they are directed towards professionals, cf. above), whereas the accompanying material is clearly different from standard academic knowledge in its appearance.

4.1 Knowledge mediation processes in HBR's digital articles

4.1.1 *The knowledge expansion processes*

In the explored articles, *the knowledge expansion processes* are related to specific strategies of combining *core knowledge* with several types of *peripheral knowledge* that can be accessed by using the network of hyperlinks existing in the articles and around them. These multimodal texts related to peripheral knowledge are meant both to show the *necessary and/or presupposed background expertise* of the articles' authors, and to help the various audiences gain more nuanced domain-specific knowledge.

First, a series of *Read More* hyperlinks are provided in order to access clarifying texts. Then, in addition to the *Read More* hyperlinks, other similar hyperlinks are also actualized through specific words, i.e. the articles' concepts and/or relevant statements on which the reader can click in order to access other articles appearing in HBR or in other journals that contextually clarify the respective concepts and/or statements. Second, hyperlinks are also provided to *expand the core knowledge* of the articles on a time-related coordinate: in the past by exploring the HBR archive of articles focused on the same topic, in the present by accessing other present articles related to the same topic, and in the future by following the topic. The hyperlinking strategy is also used in order to direct the reader's attention to other genres containing similar *peripheral knowledge*: newspaper articles, company handbooks, books, etc. It might be necessary to add that such hyperlinks are mainly targeting the practitioners who are reading the articles because the other academics do not usually need such encouragements as they both know how/where or if they need to access more knowledge. Third, the dynamic character of *knowledge expansion processes* is interactively enhanced not only by providing hyperlinks to readers' comments, but also by giving the possibility to contribute with comments. All of the hyperlinking strategies support the possibility of developing the more traditional genre of professional journal article by combining it with other instances of the same genre (related articles) or with other genres.

It is possible in this way also to combine the traditional genre with new genres (like in the case of live comments). However, what we see is rather an evolution of the genre of professional journal article through the combination with equally traditional genres to a new, up to date genre exploring the affordances of the new medium the article appears in.

Apart from employing these hyperlinking strategies, other *peripheral knowledge* is made available through the displayed artworks that accompany the articles and symbolize the articles' topics visually. For example, the photo of a huge knot visualizes symbolically the first sentence of *The Neuroscience of Trust*: "Companies are twisting themselves into knots to empower and challenge their employees" (Zak 2017).

However, two processes of *knowledge expansion* that are common to academic research articles are absent in this context: the access to new supplementary knowledge facilitated by references inside the articles and by lists of references. None of the articles include these usual strategies of academic referencing. Such an absence is possibly motivated by the need of presenting research findings in a more simplified manner in order to make the article more accessible to other readers than those belonging to the academic community.

4.1.2 *The knowledge enhancement processes*

As means of performing *evident knowledge enhancement processes* through which new knowledge is facilitated by various dissemination formats, hyperlinking strategies should be mentioned again. First, a series of hyperlinks are offered in order to access the authors' other HBR articles. These hyperlinks are manifested through the authors' names. Second, a series of hyperlinks actualized through imperatives and rhetorical questions address the readers and give them the possibilities of enhancing their knowledge: "What to read next?" and "View more from the January-February issue." These links are not part of the articles but are embedded in the same digital context, and are actualized by the images of other artworks belonging to other articles, images of other issues' cover, etc. Placed also near the article, hyperlinks to related products (i.e. reports and books) are provided, too. Furthermore, a hyperlink to the main accompanying digital material is provided. For example, in the case of the article *Getting to Sí, Ja, Oui, Hai, and Da. How to Negotiate across Cultures* (Meyer 2015), the image of a screen shot from its accompanying video: *Getting to Yes across Cultures. What to know before your next negotiation*, based on research by Erin Meyer. Knowledge is multimodally enhanced through the access to the video as it provides animating graphics that show and explain in a more accessible manner the dynamic character of the phenomena presented in the article.

Apparent knowledge enhancement processes offer the readers the possibility to access *the same knowledge* by providing hyperlinks to the PDF format of the article and its summary. Apart from that, the hyperlinking strategy is also used for the same scope in a promotional manner by *highlighting the popularity* of a respective article in the *Harvard Business Review*: "This article also appears in: HBR's 10 Must Reads 2017" (HBR 2017). Knowledge that weakens the impact of new domain-specific knowledge communicated in the articles is offered through apparent knowledge enhancement processes that give access to *unrelated knowledge* from advertising materials. When labelling multimodal data as unrelated knowledge, it should be understood in this context as knowledge that is not related to the core knowledge of the article. For example, embedded in the middle of the research articles can be encountered announcements in which the reader is encouraged to register for webinars that have no connections with the articles' topics. Additionally, the advertising aspect becomes even stronger in imperatives such as "Give the gift of success to a friend or colleague" (HBR 2015), which are supposed to persuade the reader to attract more subscribers for *Harvard Business Review*. These announcements and advertisements are visually highlighted through colourful framing and text, and they are also refreshed in order to give the readers the possibility to register for present or future webinars even if they access an article published years ago. By interrupting multimodally the reading process, the presence of such boxes containing advertising material contributes also to a new way of accessing academic knowledge similar to the way in which other digital materials that encourage text fragmentation are read/used. However, it seems obvious that the advertising discourse disrupts the academic one diluting even more its argumentation.

Summing up, we find many of the strategies known from digital academic research publishing, but also a lack of references to actual academic work, and a tendency to enhance knowledge multimodally both by promoting the article itself in an HBR context and by inserting unrelated promotional knowledge for the supplementing purpose of exploiting the channel to potential customers.

4.2 Knowledge mediation processes in the accompanying digital materials

As mentioned above, a main digital material accompanies each of the explored research-based articles in this context. We analyze the knowledge processes of the accompanying material in their own right, i.e. we are interested in multimodal differences between processes in the article and in the accompanying material rather than seeing them as a whole. In the case of *Emotion and the Art of Negotiation* (Brooks 2015), the article is accompanied by a filmed interview with its own name, *Controlling your Emotions during a Negotiation*, in which an editor of HBR asks the article's author, Alison Wood Brooks, a series of questions related to her research findings. Thus, the viewer is provided with *core knowledge* related to the topic through *knowledge expansion processes* controlled by the interviewer's questions. The casual dialogue is introduced by a medium shot of the editor who has eye contact with the viewer while presenting the author and the topic of the interview. Afterwards, the video consists of medium and long shots of the two women engaged in the discussion and having no eye contact with the viewer who is just supposed to witness the interaction. *Knowledge expansion processes* are also multimodally facilitated through a link to the original HBR article repeated both on the video's screen at the end of the interview and under the screen.

As a recurrent strategy for making apparent *knowledge enhancement processes* possible, the usage of multimodal summaries of essential advice written in colourful bullet point format on several shots of the interview can be mentioned. *Evident knowledge enhancement processes* are made possible through hyperlinks that give access to other similar digital materials related to other topics. For example, when clicking on the rhetorical question *What to watch next*, the viewer can choose from: *Related*, *The latest*, *Most popular*, *Explainers*, *Interviews* and *Tips*. The choices are related to a wide range of parameters (topic, publication date, popularity and genre); offering such a wide range of parameters can also be considered as a strategy to address various audiences apart from researchers. Visually, eight possibilities of accessing new knowledge are provided, each of them exemplified through a screen shot from the respective video.

This strategy of facilitating *evident knowledge enhancement processes* through such hyperlinked examples is also used in connection with another type of digital material, namely the video *Getting to Yes across Cultures* that accompanies the article *Getting to Sí, Ja, Oui, Hai, and Da. How to Negotiate across Cultures* (HBR 2015). In such videos, a voice-over narrator explains research findings facilitating the *knowledge expansion process* in this way, while the viewer is presented with animated graphics on the screen. The fact that this type of digital material is also supposed to address various audiences is made clear through the way in which the explanatory strategy of communication is combined with the advising one and the continuous usage of "you" pronoun. Similar to the interview, *knowledge expansion processes* are also facilitated in the video through a link to the original HBR article repeated both on the screen and under the screen. In this case, a link to the author's book is also provided.

In the webinar *The Neuroscience of Trust: Setting Leaders up for Success* that accompanies the article *The neuroscience of trust* (HBR 2017), the types of changes that appear in the knowledge mediation processes in the transmedial transition from the research-based article to the accompanying digital material are more varied. As far as the *knowledge expansion processes* are concerned, these

Table 1. Knowledge mediation processes in HBR articles and the accompanying digital material embedded in the webpages

Types of knowledge mediation processes		**Distribution of knowledge mediation processes in the genres digitally clustered around the same research work**			
		Research article	**Interview**	**Webinar**	**Explainer video**
Knowledge expansion processes affecting the breadth of the knowledge to be mediated in each of the digital materials	Mediation of core knowledge: mediation of knowledge that directly contributes to the central research topics reflected in the article's title and abstract	In the main text of the research article where contributions to central research topics are made	In the author's answers controlled by the interviewer's questions Through hyperlinking to the original research article (on and near the screen of the video) Through hyperlinks to author's HBR publications	In the research work presentation in the beginning by the HBR representative and later on by the author In the clarifying answers of the author controlled by the interviewer's questions Through the multi-modal combination of the author's voice-over and multimodal data displayed on the screen	Through the multimodal combination of voice-over explanations and animated graphics Through hyperlinking to the original research article and the book (on and under the screen of the video embedded in the webpage)
	Mediation of peripheral knowledge: mediation of background and/or supplementary knowledge that has relevance for further reading and research related to the mediated core knowledge	Through hyperlinking to: – clarifying texts; – topic-related research articles (from past, present and future); – topic-related multimodal texts (newspaper articles, company handbooks, books, etc.). In the readers' comments In displayed artworks	Through hyperlinking to related multimodal texts (near the screen of the video embedded in the webpage)		

Table 1. (*continued*)

Types of knowledge mediation processes			Distribution of knowledge mediation processes in the genres digitally clustered around the same research work			
			Research article	**Interview**	**Webinar**	**Explainer video**
	Exclusion of expected knowledge forms		Through lack of academic style referencing specific to research articles			
Knowledge enhancement processes affecting the depth of the knowledge to be mediated in each of the digital materials	Evident enhancement of knowledge	Mediation of new types of new knowledge facilitated by various dissemination formats	Inside the article: – through hyperlinking to the authors' other research work published in HBR Outside the article: – in the topic-related accompanying digital material; – through hyperlinking to other unrelated research articles, multimodal texts (reports, books, etc.), and other HBR magazine issues	Through hyperlinking to unrelated multimodal texts (near the screen of the video embedded in the webpage)	Through hyperlinking to unrelated multimodal texts (near the screen of the video embedded in the webpage)	Through hyperlinking to unrelated multimodal texts (near the screen of the video embedded in the webpage)
	Apparent enhancement of knowledge	Mediation of repeated or unrelated knowledge facilitated by various dissemination formats	Through PDF format of the research article and its summary Knowledge from advertising materials	Through summaries of essential advice for practitioners performed by the author	Through advertising summary performed by the HBR representative	

are facilitated in several stages: first, the marketing and communication director for HBR introduces and concludes the webinar off screen; second, the author presents his research work and findings in an implicit dialogue form, and then he also enters in an explicit dialogue with the listeners by answering their questions passed on by the HBR representative. *Evident knowledge enhancement processes* are facilitated by hyperlinking to author's HBR publications via author name on the website of the webinar. The author introduces his talk specifying that it will mainly consist of parts of his articles and books and highlighting the immediate practical value of the knowledge to be communicated. Although the author is not visible while talking (only a still image of his face is provided on the screen), the distance between him and the listeners is minimized by his informal manner of addressing them: *you guys, gosh, ok, give it a shot*, etc. (HBR 2017). Once again, the informal tone is supposed to indicate the affinity for being understood by more categories of listeners apart from readers of the HBR article. Furthermore, the author explicitly states: "Importantly, because I want not just to do theory" (00:11:12). However, in the still images that appear on the screen during the talk, complicated mathematical formulas are displayed clearly addressing an academic audience, together with an explicit reference to an academic article in which the underlying knowledge was first presented. This type of references is missing in articles. The rhetorical question "What to read next" addresses the readers near the webinar's screen and gives them the possibilities of *enhancing their knowledge*. Providing an advertising summary placed under the screen facilitates *apparent knowledge enhancement processes*.

Summing up, in the accompanying digital materials we see many of the same knowledge mediation processes as in the HBR articles. The first two materials (interview and explaining animation) are short and more focused, whereas the webinar is longer and goes into more details. It also has a clear reference to an academic research article on which the mediated knowledge is based. All of them demonstrate a higher degree of demonstrated awareness of the audience, which is addressed directly in different ways.

4.3 Levels of explanatory depth

So far we have investigated the processes of supporting expansion and enhancement of knowledge as meaning making processes in three selected genre clusters at the website of the *Harvard Business Review*. This kind of analysis focuses upon the interaction between elements from different modes in the meaning making process and has as its output insights into constellations of interacting modes typical for the investigated context. This type of analysis rather leans to the formal side of the content-form dichotomy. In order to also achieve insights into more content-oriented relations between texts of a cluster and thus in order to deepen out insights into differences between articles and accompanying materials, we will now switch to studying tendencies concerning the level of explanatory depth (causal relevance, causal powers, causal relations) in the two elements of a cluster. In order to be able to present a qualitative analysis within the spatial limits of our paper, we have chosen to look at only one of the three clusters investigated above, namely the HBR article *Emotion and the Art of Negotiation* and the filmed interview with the title *Controlling your Emotions during a Negotiation*.

As the interview is with the author of the article and contains a number of references to the article, we see in this case the article as the central text and the interview as a text derived from the central text. Hence, we focus upon the topic presented in the title of the article, when we look for stretches of text giving rise to the construction of explanatory knowledge structures by the user. The full title of the article is: *Emotion and the Art of Negotiation – How to use your feelings to your advantage*. We can on this basis expect explanations of the influence of feelings on negotiation (outcome) and advice as to how such insights may be used.

Starting with the article, it contains stretches of text indicating all three levels of explanatory depth to be constructed. Especially in the introduction to the article, we find examples of *causal relevance* relations:

(1) For instance, research shows that feeling or looking anxious results in suboptimal negotiation outcomes. (p. 58)

(2) There are cases when feeling angry can lead to better outcomes. (p. 61)

In these cases, we are told that there is a relation, but we do not get any details on the consequences. However, the dominant number of explanatory text parts express *causal powers* relations. In the constellation we have found to be typical, a relation is established between emotions and characteristics of negotiation, and it is stated which aspects of one side of the relation influence or are influenced by the other. Most explanations have a form where it is stated which results the existence of an emotion have on aspects of negotiation.

(3) People experiencing anxiety made weaker first offers, responded more quickly to each move the counterpart made, and were more likely to exit negotiations early (even though their instructions clearly warned that exiting early would reduce the value they received from the negotiation). (p. 59)

(4) Angry negotiators are less accurate than neutral negotiators both in recalling their own interests and in judging other parties' interests. And angry negotiators may seek to harm or retaliate against their counterparts, even though a more cooperative approach might increase the value that both sides can claim from the negotiation. (p. 61)

In Examples (3) and (4), the effects of emotions on negotiation are presented in more detail. Hence, the explanatory knowledge structures to which such formulations will give rise are intended to be more elaborate and hence at a deeper level of explanatory depth than the ones in Examples (1) and (2).

Although most causal powers structures present effects on negotiation from emotions, we also see a few examples where the direction is from specific negotiation strategies to the emergence of specific emotions:

(5) Building rapport before, during, and after a negotiation can reduce the odds that the other party will become angry. If you seek to frame the negotiation cooperatively – to make it clear that you're seeking a win-win solution instead of trying to get the lion's share of a fixed pie – you may limit the other party's perception that an angry grab for value will work well. (p. 61)

In the cases represented in Examples (3)–(5), the functional role of one factor on another is given in some detail. But there is no attempt to present what mechanisms or functional systems underlie the effect. Explanations at the latter level of explanatory depth, *causal relations*, are quite rare. They are limited to the presentation of the emotions treated in the article as an interrelated system:

(6) Anxiety is a state of distress in reaction to threatening stimuli, particularly novel situations that have the potential for undesirable outcomes. In contrast to anger, which motivates people to escalate conflict (the "fight" part of the fight-or-flight response), anxiety trips the "flight" switch and makes people want to exit the scene. Because patience and persistence are often desirable when negotiating, the urge to exit quickly is counterproductive. (p. 59)

Here we are told details about the emotion of anxiety, its (contrastive) relation to anger, and that both emotions are linked to a system of reactions, the outcome of which has a specified negative influence on negotiation. Apart from these instances, which introduce the sections treating each of the emotions in connection with negotiation in the article, all other explanations stay at the level of causal powers.

These findings demonstrate that the focus in the article is upon presenting specific influences of emotions on the quality of negotiations. More system and mechanics-oriented insights are also presented; hence the author wants the user to construct explanations at this level, too. But they are fairly scarce and obviously not the central message with the article (stretches of text indicating causal powers cover 6.2% of the text, while sections indicating causal relations 2.6%).

We have thus found indicators, connected with the explanatory complexity, that the interest in the article is in the practically useful results of academic work, as well as that academic discourse in Hyland's sense plays a central role as background and as backing in the article (see Table 2).

Table 2. Indicators of practical and academic orientation

Indicators of practical orientation	Indicators of academic orientation
Predominance of explanations at the level of causal powers	Causal relations explanations are given to all the emotions where research has been carried out
Around 18% of the text is about how to control emotions in negotiation and about examples from trainings	It is mentioned repeatedly that the causal powers explanations are based upon research (albeit without specifying fully what research)
Around 5.5% of the text is about people's experiences and folk theories about negotiation	In three cases (covering 4.5% of the text), experiments underlying the explanations are presented in some detail

As a next step, we will analyze the interview (*Controlling your Emotions during a Negotiation*) derived from the article. Like in the article, we find a dominance of explanations at the level of causal powers (consequences of emotions on negotiation) and a dominance of causal relevance structures at the beginning. The first minute of the interview is about the different emotions treated in the article (with explicit reference to the article), presenting their causal relevance for negotiation success, followed by causal power explanations for two of the emotions, anxiety and anger. Minute 2 is a recount of an example from the article of an exercise from the teaching practice of the author (without reference to the article though), followed by further causal power explanations about anger and efficient negotiation (minute 3). The last three minutes are about advice given by the author in negotiation training concerning controlling emotions. In the interview, we find no attempt to approach the level of causal relations presenting underlying mechanisms or systems. Together with the time distribution (3/6 of the time spent on advice given in training towards control of emotions in negotiation), this indicates a more pointed focus towards the interests of practitioners in achieving better practical skills based on the insights generated from the academic research, but without a focus upon this research. This means that no attempt is made to present the advice given in the interview as relying specifically upon something different from the personal expertise of the author.

However, academic research has some importance in the derived text, too, although way less than in the article:

- The author is presented as assistant professor at Harvard Business School.
- There are references to her work as an academic teacher at Harvard Business School by way of the example of a teaching exercise.
- There are three references in the dialogue to the fact that the interview is about what has been published in an HBR article (+ references from the website and in writing in the video at the end, as indicated above).

- There is one reference to an 'experiment' and hence to research in the interview (although it turns out to be an exercise in class rather than a controlled experiment).

Summing up, the characteristic of the author being a researcher and an HBR author is the central backing for the authority of her advice, not the specific results of her research into the field.

Hence, looking at the two texts from the point of view of their knowledge communication roles, we see a division of labour between the two texts:

- The interview focuses more upon the practical relevance of the academically generated knowledge, only indicates its academic character and tends to stay at the explanatory level of causal relevance and causal powers and to give advice based upon the authority of the author, without recourse to research. The interview is intended to attract attention to the article and to give practical advice (and promote instruction), rather than to present the underlying research-generated knowledge. Concerning this latter aspect, it mainly works as a teaser.
- The article in HBR has clearer indications of the academic nature of the knowledge presented and is also more clearly marked by the explanatory level of causal powers and causal relations, i.e. by deeper levels of explanation. It intends to attract attention to the underlying knowledge, or at least to the academic character of the knowledge underlying the suggested training. However, it does not invite the reader to discuss the academically generated knowledge.

5. Discussion and conclusion

We saw in Section 1 that *Harvard Business Review* considers its own mission as that of building a bridge between academia and the needs of professionals and enterprises. Based upon our analyses so far, we can now set up some hypotheses to be tested in future research on how such a mission influences the way academic knowledge is presented multimodally in the context of the hbr.org website:

- From the point of view of knowledge expansion, the task of conveying core knowledge and some peripheral background knowledge is central.
- From the point of view of knowledge enhancement strategies, references to the academic research background of the knowledge are given especially in the articles; however, these references are quite unspecific and indirect, and importantly they do not signal the articles as part of academic discourse as defined by Hyland. Especially, no citations or list of references are given.

- Specific references are given in the digital context, but mainly to other HBR articles and material, focusing upon the HBR world itself rather than upon the academic discipline of management studies.
- Some apparent knowledge enhancement strategies are solely promotional and not guided towards expanding and/or enhancing the insights of the user in the core knowledge. They are reflections of the more market-oriented purposes of HBR.
- Other apparent knowledge enhancement strategies are oriented towards making the knowledge presented more accessible in a multimodal way, for instance through repetition of the oral information in bullet points.
- The studied material accompanying the articles is basically instructional in its design, albeit aiming at different levels of detail. The differences we found in our analyses between this material and the articles are mainly due to these differences in function, rather than to a difference in the intended user of the articles and the material: Business practitioners are central receivers of all material in the website. Hence, from the point of view of stratification and diastratic variation mentioned in the introduction, we see no attempt to cover more strata.

All in all, this is in accordance with the HBR mission to function as a kind of multimodal bridge between academia and enterprises. Our analyses show that HBR intends to be independent of both sides in this multimodal bridge-building process: the articles explicitly rely upon research, but avoid important traits of academic discourse, thus indicating a distance to academia; on the other hand, even in the accompanying materials the academic status of the authors is pointed out, hence demonstrating that HBR is not a practitioners' outlet for mere experience reports. Furthermore, the fact that references are mainly to material from inside the HBR universe is probably an indication of the market orientation of this outlet. Hence, from the point of view of the focus of this volume, we can see that HBR has created a modernized genre based on the traditional genre of the professional journal by migrating it to the internet and exploiting the multimodal affordances of this medium. This enables them to include whatever generic form that may be embedded in an internet format, opening multimodal avenues for creating new types of genre clusters like the ones we have analyzed, including article + webinar and webinar + interview. However, we can also see that HBR's intent to preferably keep readers inside their own web universe generates restrictions on the genre clusters actually created. Also here, the market orientation of the universe is clearly visible.

Finally, from a methodological point of view we have here combined a procedural strategy approach typical for multimodal studies with a content-oriented approach to the study of textual basis for knowledge construction. In this way, we have been able to say not only something about the way a multimodal text intends to function as basis for knowledge construction by users, but also to formulate hypotheses about the potential results of such construction processes.

References

Bolter, Jay D., and Richard Grusin. 2002. *Remediation: Understanding New Media*. Cambridge: The MIT Press.

Engberg, Jan, and Carmen D. Maier. 2015. "Exploring the Hypermodal Communication of Academic Knowledge beyond Generic Structures." In *Discourse In and Through the Media: Recontextualizing and Reconceptualizing Expert Discourse*, ed. by Marina Bondi, Silvia Cacchiani, and Davide Mazzi, 46–65. Newcastle upon Tyne: Cambridge Scholars Publishing.

Engberg, Jan, and Carmen D. Maier. 2019. "Researchers' Move from Page to Screen: Addressing the Effects of the Video Article Format upon Academic User Engagement and Knowledge Building Processes." In *Engagement in Professional Genres: Deference and Disclosure* ed. by Carmen Sancho Guinda, 179–195. Amsterdam: John Benjamins.

Hoffmann, Lothar. 1985. *Kommunikationsmittel Fachsprache: eine Einführung*. 2 ed, *Forum für Fachsprachenforschung*, 1. Tübingen: Narr.

Hyland, Ken. 2009. *Academic Discourse: English in a Global Context*. London/New York: Continuum.

Iedema, Rick. 2003. "Multimodality, Resemiotization: Extending the Analysis of Discourse as Multi-semiotic Practice." *Visual Communication* 2 (1): 29–57. https://doi.org/10.1177/1470357203002001751

Kalverkämper, Hartwig. 1990. „Gemeinsprache und Fachsprachen – ein Plädoyer für eine integrierende Sichtweise." I *Deutsche Gegenwartsprache. Jahrbuch des Instituts für deutsche Sprache 1989*, ed. by Gerhard Stickel, 88–133. Berlin / New York: de Gruyter.

Keil, Frank C. 2003. "Categorisation, causation, and the limits of understanding." *Language and Cognitive Processes* 18 (5–6): 663–692. https://doi.org/10.1080/01690960344000062

Keil, Frank C. 2011. "The Problem of Partial Understanding." In *Current Trends in LSP Research. Aims and Methods*, ed. by Margrethe Petersen, and Jan Engberg, 251–274. Frankfurt a.M.: Peter Lang.

Kress, Gunther. 2010. *Multimodality: A Social Semiotic Approach to Contemporary Communication*. London: Routledge.

Kress, Gunther, and Theo Van Leeuwen. 2001. *Multimodal Discourse: The Modes and Media of Contemporary Communication*. London: Oxford University Press.

Maier, Carmen D., and Jan Engberg. 2013. "Tendencies in the Multimodal Evolution of Narrator's Types and Roles in Research Genres." In *Narratives in Academic and Professional Genres*, ed. by Maurizio Gotti, and Carmen Sancho Guinda, 149–173. Bern: Peter Lang.

Maier, Carmen D., and Jan Engberg. 2014. "Tendencies of Multimodal Gradations in Academic Genres Network." In *Reflections upon Genre. Encounters between Literature, Knowledge and Emerging Communicative Conventions*, ed. by Jan Engberg, Carmen D. Maier, and Ole Togeby, 113–145. Tübingen: Narr.

Maier, Carmen D., and Jan Engberg. 2016. "Challenges in the New Multimodal Environment of Research Genres: What Future do Articles of the Future Promise Us?" In *Genre Studies around the Globe. Beyond the Three Traditions*, ed. by Natascha Artemeva, and Aviva Freedman, 225–250. Edmonton: Inkshed Publications.
Norris, Sigrid, and Carmen D. Maier. 2014. "Concluding remarks." In *Interactions, Images and Texts. A Reader in Multimodality*, ed. by Sigrid Norris, and Carmen D. Maier, 381–382. Boston / Berlin: de Gruyter.
Picton, Aurélie, and Pascaline Dury. 2017. "Diastratic Variation in Language for Specific Purposes." In *Multiple Perspectives on Terminological Variation*, ed. by Patrick Drouin, Aline Francœur, John Humbley, and Aurélie Picton, 57–80. Amsterdam: John Benjamins. https://doi.org/10.1075/tlrp.18.03pic
Rozenblit, Leonid, and Frank C. Keil. 2002. "The Misunderstood Limits of Folk Science: An illusion of Explanatory Depth." *Cognitive Science* 26: 521–562. https://doi.org/10.1207/s15516709cog2605_1

Appendix: Material

Meyer, Erin. 2015. "Getting to Sí, Ja, Oui, Hai, and Da. How to Negotiate across Cultures." *Harvard Business Review* 93 (11): 74–80.
Brooks, Alison W. 2015. "Emotions and the Art of Negotiation." *Harvard Business Review* 93 (12): 57–64.
Zak, Paul J. 2017. "The Neuroscience of Trust." *Harvard Business Review* 95 (1): 84–90.
Interview. Controlling Your Emotions During a Negotiation. Accessed December 20, 2015. https://hbr.org/video/4650021133001/controlling-your-emotions-during-a-negotiation
Video with graphics & explanatory voice-over. Accessed December 20, 2015. https://hbr.org/video/4631884629001/getting-to-yes-across-cultures
Webinar. Neuroscience of Trust. Setting Leaders up for Success. Accessed December 20, 2017. https://hbr.org/webinar/2016/11/neuroscience-of-trust-setting-leaders-up-for-success

CHAPTER 8

The role of genre hybridity and hypermodality in digital knowledge dissemination

The case of the *IEEE Spectrum*

Ivana Mirović, Vesna Bogdanović and Vesna Bulatović
University of Novi Sad

New forms of discourse which have emerged as a result of synergies between traditional and digital genres utilize the resources of multimodal and hypertextual communication to change the ways in which knowledge is communicated. This case study examines the online edition of the *IEEE Spectrum* magazine, which uses the potential of new media and integrates a variety of genres to inform the members of IEEE, the world's largest engineering association, about advancements in technology, engineering, and science. The study seeks to explore in what ways the newly evolved genre of the hypermodal digital article is related to the genres traditionally used for the dissemination of scientific knowledge and how the affordances of digital format maximize knowledge dissemination.

Keywords: *IEEE Spectrum*, multimodality, knowledge dissemination, hypermodal digital article, genre hybridity, hypertextuality, recontextualization

1. Introduction

The proliferation of new media technologies has changed the ways in which scientific information is communicated. New digital environments provide the users with new experience in knowledge communication, where visualization and hypermedia play an increasingly important role (see Harmon this volume; Luzón and Pérez-Llantada this volume; Maier and Engberg this volume). Formal scholarly communication still utilizes well-established, peer-reviewed formats, such as articles in high-impact journals, monographs or conference proceedings. However, digitization has encouraged changes in the existing formats, both in the way in which knowledge is communicated on the part of the writers and in the way it is received on the part of the readers (e.g. Elsevier's *Article of the Future* project, Zudilova-Seinstra et al. 2012; Harmon this volume). The potential affordances of

https://doi.org/10.1075/pbns.308.08mir
© 2019 John Benjamins Publishing Company

digital modes of communication and new digital genres are being increasingly utilized in the process of knowledge dissemination between experts and non-experts. The use of new technologies and, in particular, new forms of multimodality, allow the "translation" of exclusive expertise into knowledge that is more suitable to the background of the addressee (Bondi, Cacchiani, and Mazzi 2015: 2; see also Maier and Engberg this volume; Orpin this volume).

In addition to multimodality, communication through new digital genres is also characterized by hypertextuality. The hypertextual nature of digital texts allows the initial information to be extended to other information sources and permits different readings of the same original text. The presence of hypertextual linkages and the use of multiple semiotic modes and media which go beyond the default conventions of multimodal genres lead to the "conflation of multimodality and hypertextuality" (Lemke 2002: 301). Lemke uses the term *hypermodality* to describe the new hyperlinked interactions between textual units, visual elements, and sound units. This term will also be used in the present paper to differentiate digitally enhanced, hypertextual and multimodal texts from plain texts.

With the technological affordances of digital media, which allow new genres to be easily incorporated into the original genre, hypermodal digital texts are frequently also characterized by genre hybridity (Crowston and Williams 2000; Luzón and Pérez-Llantada this volume). Although genre hybridity is not an exclusive characteristic of digital genres, the inclusion of characteristics associated with other genres is facilitated here by using hyperlinks or embedding texts or conventions of other genres. The digitally-enabled use of multiple semiotic modes and media, the hypertextual nature of digital communication, and the hybridity of digital genres are features which have the potential to enhance the communication of scientific knowledge (see Maier and Engberg this volume), introducing simultaneously changes in the genres which are traditionally used for the exchange and dissemination of knowledge.

This study examines the articles on the *IEEE Spectrum* website as an example of a newly developed digital genre. The *IEEE Spectrum* is a product of the Institute of Electrical and Electronics Engineers (IEEE), the world's largest association of technical professionals that publishes over 100 peer-reviewed journals. In contrast to the association's highly-specialized and much-cited professional journals, the *IEEE Spectrum* aims to inform the IEEE members (and anyone else who is interested) about the major trends, new developments and interesting stories related to the areas of technology, engineering, and science "in clear, concise language for the non specialist" (*IEEE Spectrum* Magazine *Aims and Scope*).

The website's audience can therefore be described as people who are interested in technology but are not experts in all the specific engineering areas the magazine covers. As is the case with other popular science magazines, this potential audience may include readers with various levels of expertise and can thus be described as a highly diverse group of semi-experts spanning from technology enthusiasts to professional engineers. The *IEEE Spectrum* magazine was first published in 1964, but its present format reflects the transformations which have happened in communication in the digital age, as it presents information in a synergy of traditional and digital forms. In addition to the information that would be available on the printed page of a traditional popular science magazine, the *IEEE Spectrum* articles rely heavily on communicating information through hyperlinks to other articles or websites, embedded or hyperlinked videos, animations, slideshows, gifs, etc. The statement from the website reflects the editors' awareness of the affordances offered by the digital genres: "Our blogs, podcasts, news and features stories, videos and interactive infographics engage our visitors with clear explanations about emerging concepts and developments" (*IEEE Spectrum* Magazine *About Us*). The articles featured on this website may be viewed as an independent new genre: the hypermodal digital article. The term "hypermodal digital article" will be used in this paper to refer to a digitally enhanced scientific magazine article which displays multimodal and hypertextual features. This type of article conveys information through a dynamic integration of different genres and different semiotic modes. In this hybrid genre, a number of genres are brought together for the purpose of enhancing the process of knowledge dissemination to the potential audience, as will be shown in our analysis.

The aim of the present study is to analyze the hybrid and hypermodal character of these articles. The study seeks to examine how the integration of different genres utilized in these articles and other affordances of a digitally mediated text are used to accomplish knowledge dissemination to a large and diverse audience. At the same time, the study tries to identify in what ways the newly evolved genre of hypermodal article is related to old genres, that is, genres traditionally used in the dissemination of scientific knowledge.

The following research questions are addressed:

1. What different genres can be identified within the new hybrid genre of the hypermodal article? How is this integration achieved? What role do the different genres integrated into the *IEEE Spectrum*'s articles play in knowledge dissemination?
2. How does the process of knowledge dissemination in the *IEEE Spectrum*'s hypermodal digital articles differ from the one achieved using traditional popular science articles? In what ways do the affordances of digital articles, in particular, their hypertextuality and multimodality, affect the knowledge communication process and contribute to the process of knowledge dissemination to a wide and diverse audience?

In answering these questions, our case study intends to provide insight into the practice of knowledge dissemination achieved by harnessing the affordances of digital technology. It will exemplify how the hypertextual and multimodal features of the *IEEE Spectrum*'s digital articles contribute to the knowledge building process by enhancing and expanding (Engberg and Maier 2015; see also Maier and Engberg this volume) the knowledge traditionally provided by popular science articles and it will try to demonstrate how expert technical knowledge is recontextualized using different genres in this hybrid digital format.

2. Literature review

The advances in technology and the widespread availability of digital media have resulted in the evolution, innovation and change in traditional genres (Pérez-Llantada 2016; Luzón and Pérez-Llantada this volume). This process is also perceived in the transformation of traditional popular science magazine articles into a newly emerged genre of hypermodal digital articles used on the *IEEE Spectrum* website. Although the newly emerged genre resembles its offline predecessor, "the medium adds unique properties to the web genre," such as multi-medianess and hypertextuality (Askehave and Nielsen 2005: 125). While the textual interface visuality of a digital article respects the linearity of the argument associated with the traditional written genres, selecting between multiple modes and clicking on different hyperlinks enables the reader to communicate with the text following personal interests, background knowledge or mode preferences. The analysis of digital communication has to recognize this interactivity and multimodality which greatly influence the knowledge dissemination process within the new genre. This process is accomplished in a different way from the one associated with the traditional popular science articles, where information is presented linearly from the introduction to the conclusion.

The notions of genre hybridity, hypertextuality, multimodality, and recontextualization, which are pivotal for understanding the process of knowledge dissemination within the new digital genre, are briefly discussed in the following sections.

2.1 Genre hybridity

The dynamic, hybrid character of digitally-enabled communication seems to be particularly relevant for the process of knowledge dissemination in the *IEEE Spectrum* magazine in which different genres (e.g. magazine articles, homepages, videos, gifs, animations, etc.) can be linked to achieve a common goal. Two types of genre interactions, genre embedding, when features from other texts or conventions of other genres are incorporated in the existing document, and genre linking, made possible by the presence of hyperlinks introducing another genre in the existing text (Luzón 2017), are facilitated in the digital environment as is the case of the *IEEE Spectrum* articles.

The relations created by these genres can perhaps be best described by the term genre ecology (Spinuzzi and Zachry 2000). A group of genres interrelated in a genre ecology is used to jointly mediate a given activity and this concept can be applied in the analysis of hypermodal digital articles in the *IEEE Spectrum* magazine. Genre ecologies exhibit contingency or spontaneous coordination that provides people with the opportunity to successfully accomplish a goal using diverse genres. They also exhibit decentralization or interdependency between diverse genres in mediating a given activity. In addition, genre ecologies exhibit relative stability, meaning that, even though genres are dynamic and changeable over time, new genres will tend to be interconnected in order to become conventional and consistent. These properties of genre ecologies are relevant for our analysis of the *IEEE Spectrum* articles, where a dynamic ecology of genres is applied to achieve a complex task of communicating information on new technological developments to a wide and diverse audience.

2.2 Multimodality

As modern digital communication is characterized by increased multimodality, the analysis of digital genres requires a multimodal perspective. Jewitt (2009) proposes that multimodality is a manner to approach representation, communication and interaction as something more than language, while modalities, or modes, are "semiotic resources for making meaning that are employed in a culture, such as image, writing, gesture, gaze, speech, posture" (Jewitt 2009: 1). In the multimodal approach, language is only one of many resources, i.e. modes for making meaning, occurring in a number of different media, i.e. "text or text-like entities" (Kress 2011: 38).

Communication is always multimodal but with the advent of digital technologies the use of new semiotic modes such as interactive graphics, videos, blogs, memes, podcasts and slideshows has opened the way for new forms of multimodal communication (Eriksson and Sørensen 2012; Riboni 2014; Wiggins and Bowers 2015). These new modes create new types of intersemiotic relationships. In that respect, we can examine how the resources of the written text can be interconnected with the new, digitally-enabled resources which were not available in traditional science communication. This study will analyze selected hypermodal digital articles on the *IEEE Spectrum* website and the multiplicity of modes utilized there focusing in particular on the new digital forms of communication such as animations, gifs and videos and the intersemiotic relationships between these forms and textual information. These new forms carry their own affordances, i.e. different potentials for making meaning which can be utilized for conveying information through exciting and inviting novel interactive genres (Askehave and Nielsen 2005; Zudilova-Seinstra et al. 2012; Engberg and Maier 2015). As an example of a new digital genre, the multimodal digital articles of the *IEEE Spectrum* magazine exemplify how specific affordances of multiple digital modes concur in the meaning-making process. Our analysis will show the significance of their interrelationship for successful communication to the *IEEE Spectrum* audience.

2.3 Hypertextuality

By giving access to a network of digital resources (in textual or other forms) (Adami 2013), hyperlinks provide a digital text with fluidity, i.e. the ability to expand, change, and transcend its boundaries (Djonov 2007). This fluidity makes websites more attractive for readers and allows the original information to be extended or supported in a way which can facilitate communication (see Maier and Engberg this volume). This can be particularly important for understanding complex scientific concepts. It also allows for multiple ways of reading the text; by clicking the hyperlinks, the user constructs their individual reading paths or "trajectories" (Lemke 2002: 300). The links incorporated in a digital magazine article can change the article with every click, enabling the reader to navigate and access numerous documents and sites and potentially construct the path that works best for them.

In an attempt to organize the complexity of websites as fluid hypermedia texts, Djonov (2007) proposes a website hierarchy and distinguishes website-internal and website-external hyperlinks, depending on whether they belong to the same website or they transcend and blur the website boundaries. As this classification of hyperlinks can aid in understanding the process of interconnections between different genres, it will also be used in the present analysis.

2.4 Recontextualization

Recontextualization, defined as "a process by which some part of discourse is extracted from one communicative context and conveyed into another" (Bondi et al. 2015: 2), is frequently discussed in relation to digital genres. One of the reasons for recontextualization is the need to reach a different intended audience, which can be studied in the articles from the *IEEE Spectrum* magazine. Different strategies may be used to recontextualize scientific discourse and make it accessible to a wider audience (Luzón 2013; Rowley-Jolivet and Carter-Thomas this volume). In the process of recontextualization, the elements of highly specialized discourse are subject to textual change through processes such as simplification or condensation, and technical information is refocused to stress its practical relevance. In the context of digital communication, the technological capabilities of digital genres may facilitate the recontextualization process by enabling more informal or personal interactions in the communication between experts and non-experts. This aspect of recontextualization can be observed with the *IEEE Spectrum* magazine, as exemplified later on.

3. Our study

This case study examines nine digital articles on the web page of the *IEEE Spectrum* with the aim of investigating how the affordances of knowledge communication in the digital age are harnessed in a new genre: the hypermodal digital article. The authors examined the articles in the magazine section of the *IEEE Spectrum* website in the period April-November 2018. The close analysis of the articles revealed the extensive use of new, digitally-enabled forms of communication and their integration within the text of the article, as well as regular use of hyperlinks in the text of the articles. The subsequent analysis of the hypermodal and hybrid nature of these articles focused on the distinctive characteristics of digital articles and the affordances of the new genre. The nine digital articles (DA) included in the present analysis present a purposive sample which is considered by the authors to be indicative of the various ways in which different genres and modes are combined in this new hybrid genre to aid the process of knowledge dissemination. The articles included in the analysis are listed in the Appendix.

The analysis of the articles from the *IEEE Spectrum* website seems meaningful as they can be seen as an example of a novel form of communication which develops as a result of the available technological resources. This new form of communication requires a new form of literacy, "multiliteracy" (Engberg and Maier 2015), which includes both multimodal and multimedia literacy (Jewitt and Kress 2003; Hobbs 2006; Serafini 2013). In that respect, the authors as well as the audience of the *IEEE Spectrum* can be described as multiliterate and can be viewed as pioneers or early adopters of the new form of literacy. The multimodal and hypertextual communication of the *IEEE Spectrum* can be viewed as a potential indicator of the future trends in knowledge communication or even in online magazine formats in general.

In addition, a detailed analysis of one of the digital articles (DA1) is carried out to provide insights into how, through a process of recontextualization, sophisticated technical knowledge is transformed to make it suitable for the magazine's audience. In particular, this analysis aims to examine how different genres and semiotic resources available through different multimedia formats are combined to bring this knowledge to the multitude of potential readers.

4. Methodology

The investigation into the affordances of a hypermodal digital article as a new genre in knowledge dissemination is carried out within the framework of the multileveled knowledge building model proposed by Engberg and Maier (2015; see also Maier and Engberg this volume). The model is designed to represent the knowledge building process which takes place when information is accessed through one of the articles from Elsevier's *Article of the Future* project. Engberg and Maier describe the way in which the knowledge communication process is influenced by the use of different semiotic modes and a network of hyperlinks. Consequently, their three-level model encompasses the processes on the first level, where knowledge building is made possible due to the semantic and verbal affordances of the verbal text; also, it includes the knowledge building process through multimodal interaction at the second level and through hypertextual interactivity at the third level of the model. While the multimodal knowledge building process at the second level is static and the relationships between the semiotic modes are obligatory, the presence of hyperlinks at the third level makes the relationships between different modes optional, allowing the users a choice of particular individual trajectories in the knowledge building process. The presence of the third level of knowledge building process, which allows knowledge expansion and knowledge enhancement through hyperlinks (Figure 1) is a specific characteristic of the knowledge building process through digitally mediated articles, as Maier and Engberg (this volume) also claim.

Figure 1. Knowledge building processes through hypertextual interactivity (Engberg and Maier 2015: 52)

<table>
<tr><td rowspan="6">Knowledge building process through hypertextual interactivity (knowledge building through user's dynamic navigation options)</td><td rowspan="3">Knowledge expansion</td><td>Core knowledge</td><td>Central topics (according to the title of the article's title and abstract)</td></tr>
<tr><td rowspan="2">Peripheral knowledge</td><td>Necessary and presupposed (= part of researchers' background expertise)</td></tr>
<tr><td>Supplementary (= information items relevant for further reading and research)</td></tr>
<tr><td rowspan="3">Knowledge enhancement</td><td colspan="2">New domain-specific knowledge</td></tr>
<tr><td colspan="2">Repeated knowledge</td></tr>
<tr><td colspan="2">Other types of new knowledge (usually appearing in other (non-academic) genres)</td></tr>
</table>

The model in Figure 1 provides a useful structure to address the second research question about the affordances of digital hypermodal articles. It enables us to draw a comparison between the process of knowledge building at the first and second level which happens in traditional magazine articles and the process of knowledge building characterized by expansion and enhancement (the third level) which happens in the *IEEE Spectrum* digital articles.

5. Results and discussion

5.1 *IEEE Spectrum* website

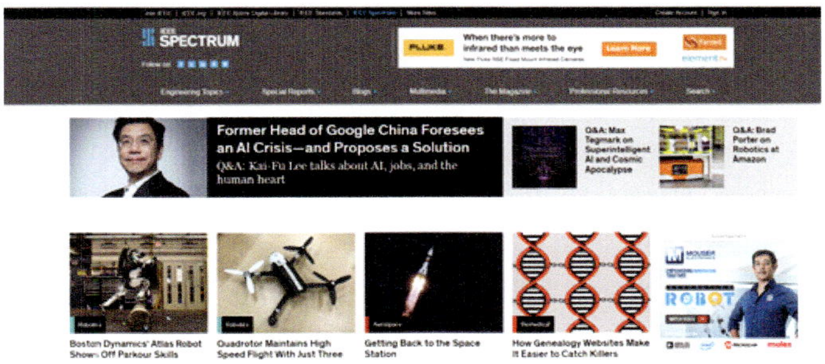

Figure 2. Top section of the homepage of the online edition of the *IEEE Spectrum* magazine (Accessed October 12, 2018)

Figure 2 shows the top section of the homepage of the *IEEE Spectrum* as seen on 24″ screen on 12th October 2018. The purpose of the homepage is threefold: (1) to present the mission and identity of the website; (2) to show the users its main parts and to preview the given information; and (3) to reveal its structure and navigation options (Krug 2000; Reiss 2000; Nielsen and Tahir 2002; Djonov 2007).

The interactive signs which are placed on the top header redirect the users to six related IEEE sites and enable the users to create an account or sign in through a click. Online interactivity is activated through clicks or touches on the screen. On the upper right side of the banner, one interactive sign redirects the user to the advertisement of the technology-related products. Below the IEEE logo are four social network symbols (*Facebook*, *Twitter*, *LinkedIn*, *Google+)*, which direct the user to the *IEEE Spectrum's* corresponding social media. By clicking on the fifth icon (*RSS feed*), the user is redirected to the *RSS feed* where he/she can subscribe to the *Spectrum* engineering topics and blog feeds. The lower section of the banner presents the menu bar with seven interactive signs (*Engineering Topics*, *Special Reports*, *Blogs*, *Multimedia*, *The Magazine*, *Professional Resources* and *Search)* which organize the website content under the appropriate topics. The click on them produces textual changes and redirects the user to the related pages. Finally, below the banner, occupying the central part of the screen, interactive signs redirect the user to the current *IEEE Spectrum* articles. The featured articles can be accessed by clicking on an article title, lead, image, video or a gif. As Djonov (2007) notices, these surfacing links allow the user to access the latest content directly from the homepage, skipping the levels in the website structure.

Almost all signs on the homepage are website-internal, i.e. the information is related to the IEEE organization. However, two interactive signs are website-external; they show greater openness of the website (sign on the menu bar – *Professional Resources* and interactive sign for advertisement) by redirecting the user to the third party websites which are not related to the IEEE. The homepage is dense with different interactive signs (images, gifs, symbols, links, text, dynamic elements) which enable users to access the presented material through a click or a touch on the screen. The website incorporates a multitude of different digital genres. In addition to the hypermodal articles which are the focus of our investigation, it includes blogs, the multimedia section with podcasts, slideshows and videos, the so-called special reports, and the professional resources section which offers webinars and whitepapers.

5.2 *IEEE Spectrum* digital hypermodal articles

The analysis of the digitally-mediated articles on the *IEEE Spectrum* website demonstrates the use of different semiotic modes and a synergy of traditional and digital genres. Besides the traditional elements of popular science magazine articles (textual material and the accompanying visuals in the form of photographs, drawings or graphs), the *IEEE Spectrum* articles also feature videos, infographics or gifs whose varied affordances are utilized to maximize the dissemination of knowledge. As a result, these articles are invariably multimodal and hybrid in nature, with different genres embedded or hyperlinked in the original text.

Although all the articles on the *IEEE Spectrum* website are characterized by the presence of numerous hyperlinks and the majority of them also feature additional embedded genres in order to provide a clearer overview of the multiplicity of genres used in the articles, the following analysis will treat embedded and hyperlinked genres separately.

5.2.1 *Embedded genres*

Our analysis of the *IEEE Spectrum* articles identified different digital genres incorporated in the textual articles by way of embedding: videos, animations, interactive maps, slideshows and gifs.

Videos are frequently embedded in the articles for the purpose of recontextualization of expert knowledge. León and Bourk (2018) stress the role of videos in multimedia texts, particularly their ability to achieve greater efficacy, involve the viewer and promote engagement in scientific issues. They also point out that science online video has a multitude of different formats and genres that are difficult to classify. We will attempt here to identify the types of videos we found embedded in the *IEEE Spectrum* articles.

A frequently found type are videos with animations or short film demonstrations which use the affordances of a different semiotic mode to clarify the information presented in the text of the article. A good example of the conscious use of the affordances of a video can be found in the article *The Troubled Quest for the Superconducting Wind Turbine* (DA2). In the text of the article, the paragraph which explains a revolutionary magnetic gearing system is introduced with the statement: "The system is hard to fully grasp unless you see it in motion..." (DA2); however, since the article also features a four-minute animated video explaining this system, the intersemiotic relationship of different modes here facilitates knowledge dissemination.

Another category of videos are those which can be interpreted as a miniature version of the traditional genre, popular science programs on TV. These videos, usually a couple of minutes long, can function independently or as part of the text they are embedded in. Following the moves of the already established TV genre, they frequently feature a presenter who introduces the topic, provides illustrations of the problem and then typically visits the site (e.g. a laboratory) where they talk to the experts in the field before concluding by presenting some practical and usually promising applications as is the case in the video embedded in the article *The Green Promise of Vertical Farms* (DA3).

While this type of video emulates a traditional TV genre, another video, such as the one found in the article *The Struggle to Make Diesel-Guzzling Cargo Ships Greener* (DA4), which incorporates an interactive map, makes full use of the unique affordances of the genre's digital format to show the movement of merchant ships and the related statistics. It allows readers to interact with the information on the screen during the actual audio presentation and to choose which different aspects of the presented information (e.g. ports, routes, ships) they want to focus on. They can also pan and zoom into the map and skip back and forward in time within the given year. As this type of interactive map represents a new form of communication, it is also accompanied by a set of questions and answers which explain how the map works and provide background information. This may be interpreted within the notion of new forms of literacy or "multiliteracy" (Engberg and Maier 2015) required to produce and read new digital genres.

In addition to videos, we frequently see animated gifs embedded in the text for the purpose of knowledge enhancement (Engberg and Maier 2015; Maier and Engberg this volume), e.g. to explain a quantum drum vibrating and not vibrating at the same time (DA5), a phenomenon which would be more difficult to understand without an artistically-rendered animation.

The articles also instantiate the presence of more than one digital genre embedded in the text. For example, a text about using robots on Mars (DA6) contains both a NASA produced video that demonstrates the operation of robots and an IEEE produced slideshow which describes graphically the different phases in their intended utilization on Mars in creating water, oxygen and fuel. The article thus uses the potentials of different modes and different genres to recontextualize the complex technical discourse and convey the ideas to its audience. Similarly, another article (DA7) starts with a video and then uses embedded gifs to enhance the reader's understanding of an innovative technology for saving water.

The above examples thus illustrate the hybrid nature of these hypermodal digital articles in which different genres are embedded within the matrix of a scientific magazine article to convey complex ideas to the diverse audience. The purpose of these genre ecologies can be interpreted within Engberg and Maier model of knowledge building process as knowledge enhancement (Engberg and Maier 2015; Maier and Engberg this volume), achieved here by using different semiotic modes and communicating through different genres.

5.2.2 *Hyperlinked genres*

The affordances provided by the hypertext character of digital articles are exploited constantly in the *IEEE Spectrum*'s articles through an abundance of site-internal and external hyperlinks. With the website's focus on the set of interrelated engineering topics concerning new technological developments, the hyperlinks frequently provide links to the other articles or videos on the *IEEE Spectrum* website discussing these topics. At the same time, the article hyperlinks also link users to external sites, such as websites of companies or institutions, homepages of the researchers or authors mentioned in the original text, company reports, *Wikipedia* pages, discussion forums or even tweets. The presence of hyperlinks in the *IEEE Spectrum* articles creates fluid and dynamic relations between different genres used in this genre ecology. The purpose of these links in the context of the knowledge building process (Engberg and Maier 2015) can be various. Most often they serve for knowledge expansion by providing wider coverage of the topics, e.g. a link to Google Patent page concerning a particular patent mentioned in the context of the article about photocells (DA8). Links to other *IEEE Spectrum* articles or research institutions can also be described in the context of knowledge expansion.

The use of infographics for the purpose of knowledge expansion in the text about the developments in telecommunications (DA9) illustrates how different genres can be closely related and their contingent relations used for constructing meaning. When the text mentions that 5G deployment is "picking up steam," a hyperlink on this phrase leads to the infographics which provide further visual details on the pace of this development.

5.2.3 *Analysis of a sample hypermodal digital article*

In order to explore how these different genres in the ecology are interrelated and to further explore how the functionalities of the hypermodal articles are utilized in the *IEEE Spectrum* articles to maximize knowledge dissemination, we will focus on a recent article from the *IEEE Spectrum* magazine entitled *Why your GPS Receiver Isn't Bigger than a Breadbox* (DA1). The semiotic resources included in this article are the following: written language, audio, video, color, image, spatial distribution and hyperlinks. All these elements have meaning potential and, by combining them, intersemiotic coherence is created between the modes, which enables the readers to construct meaning.

The expert knowledge about the development of GPS is recontextualized in this article to focus on a researcher who was one of the key figures in the development of the GPS system. The information is presented through a hypermodal article which takes the central part of the screen and is accompanied by a short CV of the researcher, Brad Parkinson. The familiar icons on the left side of the screen enable the reader to easily share the article on the social networks, while *Editor's picks* on the right (see Figure 3) suggests other *IEEE Spectrum* articles on the same topic which readers can access if they want to extend their knowledge of GPS.

The story is presented in the form of a digital article with numerous hyperlinks and, in addition to the traditionally found visual elements such as photos, it also includes an embedded video (see the image in Figure 3).

Figure 3. A screenshot of a section of the web page from the article "Why your GPS Receiver Isn't Bigger than a Breadbox" (Accessed October 12, 2018)

The video is 5–6 minutes long and mainly features Parkinson talking about his early interests and professional development, explaining his invention. There is also a narrator who takes us through Parkinson's biography, commenting and emphasizing particular twists and turns. The video also includes photos, maps, post-it notes, drawings, animations and a short extract showing Parkinson lecturing at Stanford University. The video can be interpreted as a summarized version of the article, but the information presented here is recontextualized and transformed in a way which is more engaging for the audience: Parkinson talks directly to the camera i.e. the potential viewers, there are photographs from his family album, animations showing how GPS works, examples of people using GPS in everyday life as well as a clip with a joke he shares with the students at Stanford. All these features make it suitable for an audience with little background knowledge, as the information from the article is recontextualized through the process of simplification and condensation (Bondi et al. 2015). In addition, the expert knowledge about GPS systems is here refocused to concentrate on the person who contributed to its invention and on the practical applications of the new technology instead of the technical details. The knowledge recontextualization and the transfer of information are maximized by employing video's multimedia affordances.

The article itself follows the same narrative as the video (Parkinson's professional development, which is linked to the development of GPS), but it includes more technical details, e.g.

> Parkinson invented a sensor that could tell the position of the rotor relative to the desired axis. Using an algorithm he called hemispheric torquing, he could then apply a magnetic field to adjust the rotor's position, sending it spinning along the desired axis without changing its overall position in space.
>
> (*Why your GPS Receiver Isn't Bigger than a Breadbox*)

At the same time, there are elements of the article which add emotions and sympathy in a human-interest story, e.g. the part explaining how an important breakthrough was made during one Labor Day weekend when there was nobody around at the office except Parkinson and his small team. This illustrates the tendency to combine elements of both "soft" and "hard" reporting (Reinemann et al. 2011), which is evident throughout the article. The so-called "soft" reporting focuses primarily on the people involved in or affected by a particular event and stresses their personal experience in order to engage the readers and present the information in the way which appears more relevant to a wider audience. In contrast to this focus on individual stories, "hard" reporting is more concerned with the facts and focuses on the accurate representation of information without personal involvement (Reinemann et al. 2011). The digital format of the *IEEE Spectrum* article enables the author to address the issue from both aspects: the use of multimedia and hyperlinks and the presence of different genres alternately emphasize the "soft" and "hard" elements of the development of GPS system. This can also be observed in Parkinson's short CV accompanying the main text. In addition to providing the information about his education and career, it unconventionally also informs the reader about his favorite food, restaurant, books and music. Furthermore, exploiting the affordances of the article's digital form allows the reader to immediately order Parkinson's favorite book by following a hyperlink to the Amazon website.

Undoubtedly, the greatest affordances of the article's digital form can be found in the use of hyperlinks. The hyperlinks are both internal (providing links to different multimodal resources within the *IEEE Spectrum* website) or external (providing links to information outside this website). The hypertextual interactivity contributes to the knowledge building process by *expanding* the information from the text and also by *enhancing* the understanding of the information from the main article (Engberg and Maier 2015). Hyperlinks are frequently used throughout the text for the purpose of knowledge *expansion* to provide supplementary technical knowledge about the topic of GPS. For example, when Parkinson talks about his PhD thesis and his work on free rotor gyroscope, the external hyperlink directs the reader to the original 1970 journal article where Parkinson and his supervisor first published their results and explained this concept.

The knowledge building process is also facilitated by the links which contribute to knowledge *enhancement*, such as a link to John Deere website, which illustrates how the technology that Parkinson helped develop is now used for GPS guided tractors. This digitally mediated article, with its multimodal and hyperlinked features, has affordances which advance the knowledge building process. Its potentials lie in the easy access to hypertextual resources which can complement the primary information presented in the text of the article, but also in the fact that the article's digital features facilitate the use of multiple genres (e.g. company websites, scientific journal article, military reports, discussion forum). These can present the information in an alternative way which potentially appeals to non-specialist audience, attracts their attention, and feeds their curiosity. This is potentially the case with the link to the 97-page document which was compiled while Parkinson worked for the US Army. The readers who follow this hyperlink will be in a position to glance through the original 1966 military document with handmade drawings which remained classified until 1979. This document is probably not included with the idea that it will inspire the *IEEE Spectrum*'s audience to read it, but its accessibility gives the audience a sense of satisfaction that this insight has been made available to them by a simple click on a hyperlink.

On the one hand, the potentials of the new digital genre allow a potential reader to accumulate vast knowledge about the topic of GPS through a combination of text and various hyperlinked genres. Alternatively, this expert technical knowledge is recontextualized in terms of condensing, simplifying and refocusing it for a non-expert audience. An illustration of this is the representation of Parkinson's work at Stanford University: while an article hyperlink directs the reader to the University website, which provides an abundance of information about the project on which Parkinson worked, the part of the video dealing with his engagement at Stanford University shows him sharing a joke about GPS with the students.

The fact that a digital article allows multiple information flows and individualized or "personalized" versions of the same text is particularly suitable for the *IEEE Spectrum* readership. It was mentioned earlier that their potential audience is very diverse: lay audience, technology enthusiasts, engineering students, professional engineers interested in the latest developments, or even technology experts from related fields. By exploring the potentials of different semiotic modes and the technology-based affordances, the *IEEE Spectrum* magazine offers these users the possibility to choose their individual trajectories in accessing the available information. Through this digital genre, the *IEEE Spectrum* succeeds in communicating knowledge to the audiences with different knowledge backgrounds, different purposes for reading, different levels of commitment or different ways in which they like to interact with their texts.

6. Conclusion

The study has sought to explore how the affordances of a new digital genre are utilized to accomplish successful knowledge dissemination to a large and diverse audience. It has focused on the digital articles on the *IEEE Spectrum* website in order to identify and investigate different genres they employ towards achieving this objective and the integration of different generic characteristics in a hybrid new genre. Our inquiry aimed to provide additional support for the research into the multigeneric and multimodal nature of digital communication.

The results of the case study of the *IEEE Spectrum* articles indicate that digital articles can be regarded as a specific genre, a hypermodal digital article. In this novel hybrid genre, diverse genres and semiotic modes are brought together in order to fulfill the magazine's aim. As the new resources are introduced and incorporated in the traditional article format, the characteristics of the genre are changed to integrate existing and newly developed genres. The result is a dynamic, multigeneric form in which information is communicated through verbal text, but also through many other genres and modes embedded in the text or accessible through internal or external hyperlinks.

The aim of the paper was twofold. The first research question was concerned with the hybrid nature of the new genre, different genres integrated within the hypermodal digital article and their role in accomplishing knowledge dissemination. The analysis of the *IEEE Spectrum* digital articles indicated extensive, technologically facilitated, exploitation of conventions and resources of various generic forms. In this genre ecology the purpose of knowledge dissemination is achieved by a synergy of traditional and newly developed genres. Since a multitude of different genres and modes are combined through genre mixing, embedding and linking, the resulting hybrid form is a genre which has a highly dynamic and variable nature. The new genre thus frequently features embedded gifs, slides or animations (in addition to traditional pictures or drawings), while the text's hypertextual links provide access to other articles as well as to company websites, researchers' homepages or discussion forums.

Through the process of recontextualization, parts of discourse are extracted from one communicative context and conveyed into another. In the *IEEE Spectrum* articles, the technical discourse of the verbal text is frequently recontextualized using videos which present the information in a condensed or simplified way, with the focus on the practical applications or impact on everyday life. In that process, the videos embedded in the *IEEE Spectrum* article typically exploit the conventions of an already established genre: TV programs which are traditionally used for the popularization of science. In addition, the highly technical aspects of the engineering developments presented in the text are commonly recontextualized in the form of animations or slideshows. At the same time, the digital nature of these articles and, particularly, their hypertextuality allow the appropriation of different discursive and generic resources and their interaction. In the article which was examined in this study, we identified the websites of different companies and research institutions, other *IEEE Spectrum* articles, a scientific journal article, a declassified military document, and a discussion form as the illustrations of the different genres related in a dynamic genre ecology. In any other article, a different cluster of genres may be observed in the genre ecology as they are interrelated, coordinated and adapted to accomplish the desired objectives.

The second research question attempted to identify the changes in the process of knowledge dissemination which accompany the introduction of hypermodal articles. In that respect the process of acquiring knowledge through a newly evolved digital genre is compared to the traditionally enabled dissemination of scientific knowledge in popular science magazine articles on the basis of the model proposed by Engberg and Maier (2015). Our analysis of nine sample *IEEE Spectrum* articles and a cluster of genres identified within this digital generic form demonstrated that the greatest affordance offered by the hypermodal digital article lies in the use of hyperlinks which contribute to the knowledge building process by expanding the information from the text and also by enhancing the information comprehension (Engberg and Maier 2015). A hyperlinked digital text may be particularly beneficial for knowledge expansion by providing an easily accessible link to additional information sources which offer supplementary information. In the *IEEE Spectrum* articles, a network of *Spectrum* internal and external hyperlinks provided by the article authors offers a vast amount of information from a thoroughly researched area of expert knowledge. This would be impossible to convey without harnessing the potentials of the digital article's hypertextual resources.

On the other hand, the process of knowledge enhancement is frequently facilitated by the incorporation of different forms, modes or genres such as gifs, interactive maps or animated videos. As the multimodal perspective can help us understand the process of knowledge enhancement, these become almost standard features of the *IEEE Spectrum*'s digital articles. The specific affordance of the hypermodal character of the *IEEE Spectrum* articles is that it allows users to choose individual trajectories in accessing the available information depending on their individual preferences, previous background knowledge or learning styles. This enables the information presented in the text to be contracted and expanded with each new reader and each new reading. The use of different semiotic modes which enable the readers to engage in different ways along with the ability to create one's own personalized version of the text are particularly beneficial for the *IEEE Spectrum*'s diverse potential audience.

This new digital environment uses multiple semiotic modes and media, the hypertextual nature of digital communication and the hybridity of digital genres to improve the dissemination of scientific and technical knowledge and allow a dynamic knowledge communication suitable to a wide range of experts and non-experts. The new form of communication relies on multiliteracy, a new form of literacy required both on the side of the producers and on the side of their audience(s). As digitally-enabled communication becomes a ubiquitous feature of knowledge exchange and dissemination, the newly developed genre of hypermodal articles may emerge as a standard article form. It has the potential to change future communication of knowledge by facilitating individual knowledge building processes and engaging larger potential audiences. Hence, the multimodal and hypertextual communication presented in the *IEEE Spectrum* online magazine can be an indicator of future trends in hypermodal digital genres.

References

Adami, Elisabetta. 2013. "A Social Semiotic Multimodal Analysis Framework for Website Interactivity." *NCRM Working Paper*. London: NCRM. (Unpublished). Accessed August 19, 2018. <http://eprints.ncrm.ac.uk/3074/4/website_interactivity_Adami.pdf>

Askehave, Inger, and Anne E. Nielsen. 2005. "Digital Genres: A Challenge to Traditional Genre Theory." *Information Technology & People* 18 (2): 120–141. https://doi.org/10.1108/09593840510601504

Bondi, Marina, Silvia Cacchiani, and Davide Mazzi. 2015. “Discourse In and Through the Media: Recontextualizing and Reconceptualizing Expert Discourse.” In *Discourse In and Through the Media: Recontextualizing and Reconceptualizing Expert Discourse*, ed. by Marina Bondi, Silvia Cacchiani, and Davide Mazzi, 1–21. Newcastle upon Tyne, UK: Cambridge Scholars Publishing.

Crowston, Kevin, and Marie Williams. 2000. “Reproduced and Emergent Genres of Communication on the World Wide Web.” *The Information Society* 16 (3): 201–215. https://doi.org/10.1080/01972240050133652

Djonov, Emilia. 2007. “Website Hierarchy and the Interaction between Content Organization, Webpage and Navigation Design: A Systemic Functional Hypermedia Discourse Analysis Perspective.” *The Information Design Journal* 15 (2): 144–162. https://doi.org/10.1075/idj.15.2.07djo

Engberg, Jan, and Carmen D. Maier. 2015. “Exploring the Hypermodal Communication of Academic Knowledge beyond Generic Structure.” In *Discourse In and Through the Media: Recontextualizing and Reconceptualizing Expert Discourse*, ed. by Marina Bondi, Silvia Cacchiani, and Davide Mazzi, 46–62. Newcastle upon Tyne, UK: Cambridge Scholars Publishing.

Eriksson, Thommy, and Inge E. Sørensen. 2012. “Reflections on Academic Video.” In *Seminar. net – International Journal of Media, Technology and Lifelong Learning* 8 (1). Accessed August 19, 2018. https://journals.hioa.no/index.php/seminar/article/view/2399

Hobbs, Renee. 2006. “Multiple Visions of Multimedia Literacy: Emerging Areas of Synthesis.” In *International Handbook of Literacy and Technology*, ed. by Michael McKenna, Linda Labbo, Ronald Keiffer, and D. Reinking. 2, 15–28. Mahwah, NJ: Lawrence Erlbaum Associates.

Jewitt, Carey (ed.). 2009. *Handbook of Multimodal Analysis*. London: Routledge.

Jewitt, Carey, and Gunther Kress. 2003. *Multimodal Literacy*. New York: Peter Lang.

Kress, Gunther. 2011. “Multimodal Discourse Analysis.” In *The Routledge Handbook of Discourse Analysis*, ed. by James P. Gee, and Michael Handford, 35–50. London: Routledge.

Krug, Steve. 2000. *Don’t Make Me Think: A Common Sense Approach to Web Usability*. Indianapolis, Indiana: New Riders.

Lemke, Jay L. 2002. “Travels in Hypermodality.” *Visual Communication* 1 (3): 299–325. https://doi.org/10.1177/147035720200100303

León, Bienvenido, and Michael Bourk. 2018. *Communicating Science and Technology through Online Video*. London, UK: Routledge. https://doi.org/10.4324/9781351054584

Luzón, María-José. 2013. “Public Communication of Science in Blogs: Recontextualizing Scientific Discourse for a Diversified Audience.” *Written Communication* 30 (4): 428–457. https://doi.org/10.1177/0741088313493610

Luzón, María-José. 2017. “Connecting Genres and Languages in Online Scholarly Communication: An Analysis of Research Group Blogs.” *Written Communication* 34 (4): 441–471. https://doi.org/10.1177/0741088317726298

Nielsen, Jakob, and Marie Tahir. 2002. *Homepage Usability: 50 Websites Deconstructed*. Indianapolis: New Riders Publishing.

Pérez-Llantada, Carmen. 2016. “How Is the Digital Medium Shaping Research Genres? Some Cross-Disciplinary Trends.” *ESP Today* 4 (1): 22–42.

Reiss, Eric L. 2000. *Practical Information Architecture: A Hands-on Approach to Structuring Successful Websites*. London/Harlow: Pearson Education Ltd.

Reinemann, Carsten, James Stanyer, Sebastian Scherr, and Guido Legnante. 2011. “Hard and Soft News: A Review of Concepts, Operationalizations and Key Findings.” *Journalism* 13 (2): 221–239. https://doi.org/10.1177/1464884911427803

Riboni, Giorgia. 2014. "Constructing (cyber-) Space on Twitter: A Study of Place Deixis in Tweets." In *Space, Place and the Discursive Construction of Identity*, ed. by Julia Bamford, Franca Poppi, and Davide Mazzi, 217–236. Bern: Peter Lang.

Serafini, Frank. 2013. *Reading the Visual: An Introduction to Teaching Multimodal Literacy*. New York: Teachers College Press.

Spinuzzi, Clay, and Mark Zachry. 2000. "Genre Ecologies: An Open-system Approach to Understanding and Constructing Documentation." *ACM Journal of Computer Documentation* 24 (3): 169–181. https://doi.org/10.1145/344599.344646

Wiggins, Bradley E. and G. Bret Bowers. 2015. "Memes as Genre: A Structurational Analysis of the Memescape." *New Media & Society* 17(11): 1–21. https://doi.org/10.1177/1461444814535194

Zudilova-Seinstra, Elena, Martijn Klompenhouwer, Frans Heeman, and IJsbrand J. Aalbersberg. 2012. "The Elsevier Article of the Future Project: A Novel Experience of Online Reading." In *The Future of the Academic Journal*, ed. by Bill Cope, and Angus Phillips, 357–378. Chandos Publishing series.

Appendix: List of *IEEE Spectrum* analyzed articles

DA1 *Why your GPS Receiver Isn't Bigger than a Breadbox* (https://spectrum.ieee.org/geek-life/profiles/why-your-gps-receiver-isnt-bigger-than-a-breadbox) (Published online 19/04/2018)

DA2 *The Troubled Quest for the Superconducting Wind Turbine* (https://spectrum.ieee.org/green-tech/wind/the-troubled-quest-for-the-superconducting-wind-turbine) (Published online 26/07/2018)

DA3 *The Green Promise of Vertical Farms* (https://spectrum.ieee.org/energy/environment/the-green-promise-of-verticalfarms) (Published online 02/06/2018)

DA4 *The Struggle to Make Diesel-Guzzling Cargo Ships Greener* (https://spectrum.ieee.org/transportation/marine/the-struggle-to-make-dieselguzzling-cargo-ships-greener) (Published online 29/05/2018)

DA5 *A Quantum Drum Brings Quantum Mechanics to the Macroscale* (https://spectrum.ieee.org/nanoclast/semiconductors/nanotechnology/a-quantum-drum-brings-quantum-mechanics-to-the-macroscale) (Published online 22/05/2018)

DA6 *How NASA Will Use Robots to Create Rocket Fuel from Martian Soil* (https://spectrum.ieee.org/aerospace/robotic-exploration/how-nasa-will-use-robots-to-create-rocket-fuel-from-martian-soil) (Published online 30/10/2018)

DA7 *This Innovative Technology Harvests Water from Cooling Towers* (https://spectrum.ieee.org/energywise/energy/environment/this-innovative-technology-harvests-water-from-cooling-towers). (Published online 08/06/2018)

DA8 *This Photocell Generates Both Power and Hydrogen* (https://spectrum.ieee.org/energywise/green-tech/solar/hybrid-photocell-generates-power-and-hydrogen) (Published online 01/10/2018)

DA9 *It's Never too Early to Think about 6G* (https://spectrum.ieee.org/tech-talk/telecom/wireless/its-never-too-early-to-think-about-6g) (Published online 22/05/2018)

CHAPTER 9

#Vaccineswork

Recontextualizing the content of epidemiology reports on Twitter

Deborah Orpin
Wolverhampton University

This study examines the ways in which information originating in epidemiological reports is recontextualized in the @ECDC_VPD account, the Twitter account of a European health agency. Using a corpus-assisted discourse analytical approach complemented with multimodal analysis, this study compares the strategies used to achieve *proximity* (Hyland 2010) in the space-constrained genre of Twitter with those used in the source texts. The study finds that the macro-structural properties of the @ECDC_VPD tweets have become more complex over time and the use of images to enhance meaning-making has increased. The drive to present claims as newsworthy, coupled with the 140/280-character constraint, results in the tweets containing greater relative use of stance markers and lower use of epistemic modals than is observed in the source texts. The @ECDC_VPD tweets display a greater range of engagement strategies than is seen in the source texts.

Keywords: digital genres, popularization, recontextualization, Twitter, health communication

1. Introduction

This chapter focuses on the recontextualization of scientific and health-related information in the tweets of the @ECDC_VPD Twitter account, the account belonging to the European Centre for Disease Control and Prevention's (ECDC) Vaccine Preventable Diseases programme (VPD). The main objective of the ECDC-VPD is to support EU member states in preventing and containing the spread of vaccine-preventable diseases and invasive bacterial infections (Kramarz et al. 2014: 1). To that end, it gathers, analyzes and disseminates disease surveillance and vaccine coverage data and commissions scientific reports in order to

https://doi.org/10.1075/pbns.308.09orp
© 2019 John Benjamins Publishing Company

provide robust evidence to inform policy-making (ibid.). It is also responsible for promoting vaccine acceptance among the public (ibid.:5). The @ECDC_VPD Twitter account is an important tool in the agency's communication strategy. Social network messages might not strike one immediately as a popularization genre. However, popularization concerns the recontextualization of scientific knowledge in a variety of contexts, realized through a range of discursive-semiotic practices (Calsamiglia and van Dijk 2004: 317) and involving diverse genres (Hilgartner 1990: 528). If the presentation of scientific knowledge can be said to lie on a continuum, with "laboratory shop-talk," for example, lying "upstream" of the scientific research paper and genres such as policy reports and media articles lying "downstream" (Hilgartner 1990: 528), the @ECDC_VPD tweets can be seen as lying on the downstream end of the spectrum.

The authors of the @ECDC_VPD tweets are in a similar position to writers of academic science blogs in that they must address a heterogeneous audience. The core audience comprises scientists, policy makers and healthcare professionals, while the secondary audience consists of members of the public. The writers therefore need to engage with, and accommodate to the diverse needs of, the various constituencies, by exploiting the affordances of the technology and adopting particular discursive strategies in order to render specialized knowledge accessible (Luzón 2013: 429). The @ECDC_VPD authors also have to exhort readers to action, whether encouraging a health professional to read a document or persuading members of the public to get vaccinated. In addition, they must accommodate to the limitations of Twitter's 140/280-character constraint.

Few studies to date examine relations between traditional science genres and Twitter. Exceptions are Puschmann's (2014) investigation of the ways in which academics disseminate science through digital genres, among them tweets, and the studies by Shan et al. (2013) and Büchi (2016), which consider the synergies between science reports in traditional news genres and scientific discourse on Twitter. The current study contributes to research in this field by examining the relations between the @ECDC_VPD tweets and the scientific reports which inform them. Using a corpus-assisted discourse analytical approach, it aims to discover the discursive-semiotic strategies used by the @ECDC_VPD tweet authors to recontextualize scientific information in a character-constrained genre targeted at a heterogeneous audience. The main theoretical framework informing analysis and interpretation in this study is Hyland's (2010) model of *proximity*. Analysis is enhanced by drawing on Hoey's (1983) model of discourse organization and multimodal analysis (Kress and van Leeuwen 1996). This chapter continues with a discussion of the ways in which proximity resources are exploited in traditional and digital genres, before articulating the research questions. The data and methods are then described and the implications of the findings discussed.

2. Exploiting proximity resources in traditional and digital genres

2.1 Proximity resources

Proximity refers to the strategies a writer uses to interact with the reader and tailor the ideational content of the text to their needs (Hyland 2010: 117). There are five facets of proximity: Organization, Argument, Credibility, Stance and Engagement. The first three relate to the ways in which writers shape their material and demonstrate authority through the use of academic or general writing conventions. Of the latter two, stance concerns "the ways in which writers present themselves and convey their judgements, opinions and commitments" (Hyland 2005: 176), while engagement relates to the ways in which the writer acknowledges the presence of the reader and guides them through the argument (Hyland 2010: 125). Hyland (2005: 177) points out that evaluative meaning can be expressed in a wide variety of ways and not always using implicitly evaluative language. He stresses, therefore, that his model cannot account for all types of expressions of stance and engagement but that it can describe the most frequently occurring forms.

Stance, Hyland (ibid.: 178–181) explains, is typically realized through the use of self-mentions (use of *I*, *we*, *my*, and so on), modal verbs and other hedging devices (such as *might*, *possible*, *perhaps*), boosters (words and phrases which intensify the force of a proposition, for example *clearly*, *obviously*), and attitude markers. Attitude markers are words and phrases which express the writer's affective attitude to propositions. The more explicit among these are items such as sentence adverbs (for example, *unfortunately*), adjectives (for example, *remarkable*) and "attitude verbs" (such as *agree*, *prefer*). Finally, engagement is realized through the use of reader pronouns, personal asides, appeals to shared knowledge, directives, and questions.

2.2 Organization, argument and credibility

As Hyland (2010) argues, all writers need to appeal to the reader's sense of novelty. The difference between the research article and popularization lies in how the appeal to novelty is expressed. According to Hyland (ibid.: 119–120), the author of the research article presents novelty in terms of the contribution the work makes to science whilst the popularization writer presents it in terms of newsworthiness. Thus, where the research article locates the main claim in the context of previous research and emphasizes the scientific process, the popularization foregrounds the main claim by placing it at the beginning of the article and presenting it "in terms of what is of immediate value or potential benefit to readers" (Hyland ibid: 120). Where digital popularization genres are concerned, the organizational structure depends a lot on the context. The science blog, for example, conforms to the standard popularization structure, placing the main claim at the top of the post (Luzón 2013: 452), while in TED talks, the main claim is prefaced by accounts which contextualize the topic for the audience (Caliendo 2014: 118–119; Scotto di Carlo 2014: 594). Certain structures which help signal how a reader is to interpret a message are found across diverse genres. A commonly used macro-textual structure is the Situation-Problem-Solution-Evaluation (SPSE) structure: the writer describes a situation, out of which a problem arises, and then describes and evaluates a response or solution to the problem (Hoey 1983: 43–56). Another frequent pattern is the General-Particular clause relation, in which a generalization is followed by an example, or a preview followed by detail (Hoey ibid.: 134–143).

In order to accommodate to the needs of non-specialist readers, popularization authors need to include explanatory structures. One such object is the graphical image. Graphical images play a role in organizing arguments in research articles and popularizations alike, but their function differs according to context (Hyland 2010: 119). In the research article, objects such as tables, graphs, and figures provide evidence to support the scientist's argument and to allow the expert reader to reach their own interpretation. In traditional popularizations, they are used to entice the reader or explain and simplify findings (Miller 1998: 31). How they function in the @ECDC_VPD tweets may depend on the target audience. In science blogs aimed mainly at an expert audience, figures and illustrations function much as they do in scientific journal articles (Puschmann 2014), while in health promotion literature, images typically act as visual metaphors (Maci 2014: 177), permitting the quick and easy comprehension of essential information in support of the written text (Maci 2014: 185). Whatever the genre, aspects of the visual grammar of images remain constant: given information tends to occur on the left of an image, new information on the right (Kress and van Leeuwen 1996: 186–192); information in the top half of an image represents what is "ideal," that in the lower half as what is "real" (ibid.:193–194). The use of colour, meanwhile, can express the degree of commitment to the truth of an image, much as the modal system in language does. Bright, heavily saturated colours convey an un-modalized message and muted, less saturated colours suggest modalization (ibid.:165–168).

2.3 Stance and engagement

It has long been recognized that, compared with research articles, popularizations contain relatively few modal verbs and hedging devices and make greater use of intensifying expressions (Fahnestock 1986: 282–283). This occurs because the drive to present novel information as newsworthy leads popularization writers to express claims with a greater degree of certainty than in the original article. The same is true for digital popularization genres, such as TED talks (Caliendo 2014) or blogs (Luzón 2013) (see also Rowley-Jolivet and Carter-Thomas this volume, for the case of oral genres). A higher than average occurrence of attitude markers has been also observed in tweets about science (Shan et al. 2013; Veltri and Atanasova 2015; Büchi 2016). However, this may arise because the expression of affect is a typical generic feature of Twitter interaction. Researchers have noted that Twitter interaction shows markedly frequent use of expressions of stance (Titak and Roberson 2013: 247–249), whilst modals are relatively infrequent (Knight 2015: 29).

Although no popularization study has yet examined engagement strategies on Twitter, the extent to which the @ECDC_VPD tweets are likely to show features typical of popularizations remains to be seen. In traditional popularizations, writers typically make greater use of second person pronouns and questions in order to engage the reader. However, the fact that tweets are rarely addressed to a specific individual (Knight 2015: 29) might militate against their use.

3. Research questions

The research questions guiding the study are as follows.

1. What are the typical organizational and argument structures of the @ ECDC_VPD tweets and to what extent, and how, do these structures resemble or differ from those of the source texts on which they draw?
2. How do the @ECDC_VPD authors exploit particular discursive-semiotic resources and the affordances of Twitter to overcome the challenges of space constraints and how does the use of images on the @ECDC_VPD account compare with that in the source texts?
3. Which linguistic features are most frequently used in the @ECDC_VPD tweets to realize stance and engagement strategies and to what extent and how do these resemble or differ from those used in the source texts on which they draw?

4. Data and method

4.1 Data and corpus compilation

The @ECDC_VPD Twitter account was opened in September 2010. At the time this project was conducted, tweets dating from between 5 June 2013 and September 2018 were accessible to view. Analysis of organization, argument and credibility in the @ECDC_VPD tweets is based on this data. To aid analysis of stance and engagement strategies, two corpora were compiled: the @ECDC_VPD corpus and the ECDC Reports corpus. The @ECDC_VPD corpus was compiled using the social media data analysis toolkit, FireAnt (Anthony and Hardaker 2017). Once FireAnt had gathered and processed the tweets, linguistic data was exported and saved in text file format. The @ECDC_VPD corpus consists of 1,508 tweets, including 771 retweets, dating from between 9 September 2010 and 25 July 2018, and comprises 30,481 running words.

The ECDC Reports corpus was compiled manually. Reports produced by the ECDC and to which @ECDC_VPD tweets dating from January 2017 to July 2018 contained a link were downloaded and saved in PDF format. They were then copied and pasted into text file format. The ECDC Reports corpus consists of 25 reports and comprises 216,986 running words. ECDC reports are of two types: surveillance reports, which summarize and interpret the implications of epidemiological data submitted by EU member states; and technical (or expert opinion) reports, which present the findings of research projects commissioned by the ECDC and make policy recommendations, often on the basis of evidence contained in the surveillance reports. Seventeen of the 25 reports in the ECDC Reports corpus are surveillance reports, totalling 90,822 running words; the remaining eight are technical reports, totalling 126,164 running words. Material produced by the ECDC and made publicly available may be reproduced without prior permission, provided the ECDC is cited as the source (ECDC 2005–2019).

4.2 Method

The @ECDC_VPD tweets which were accessible online were examined in order to discover how the proximity resources of organization, argument and credibility were exploited. Hoey's (1983) model was used to analyze macro-structural patterns and multimodal analysis (Kress and van Leeuwen 1996) was used to analyze the use of graphic images. A record was kept noting the target audience each tweet was aimed at (professionals or the general public), their macro-structural patterns, for example, General-Particular or Situation-Problem-Solution-Evaluation (Hoey 1983), and the presence or absence of a graphic image. Where a graphic image was included in a tweet, the type of image (map, graph, photo, and so on) was recorded. The use of proximity resources in the tweets was compared with their use in the ECDC reports.

To analyze the use of stance and engagement resources in the @ECDC_VPD and ECDC Reports corpora, AntConc (Anthony 2019) was used to provide frequency data and evidence of usage. Firstly, keyword lists were compiled for each corpus in order to identify explicit markers of stance or engagement which occur with markedly high or low relative frequency when compared with a reference corpus. The BNC was used as reference corpus and Log-likelihood (LL) as the measure of significance. One drawback to relying on keyword analysis alone is that one risks overlooking some salient phraseological patterns. In order to uncover as wide a range as possible of phraseological patterns, for each corpus, a list was compiled of two- to three-word n-grams with a minimum frequency of five occurrences. The lists were scanned to identify n-grams corresponding to, or containing, an explicit marker of stance or engagement. Where necessary, concordance evidence was consulted to establish how specific terms are used in the relevant corpus. Concordance analysis was then conducted on the keywords and n-grams identified.

5. Results and discussion

5.1 Credibility

Essentially, the credibility of the information disseminated by the ECDC and ECDC-VPD, whether through the reports they produce or through the Twitter account, is implicit in their status role as EU health agencies. However, authority is also signalled in aspects of the design and layout of materials they produce. We can see this if we look at a typical @ECDC_VPD tweet (see Figure 1). The expression of authority is reinforced by features we can see in the profile picture on the left of the screen. The bold typeface and straight lines with which the initials "VPD" are written can be interpreted as suggesting strength and solidity (Machin 2007: 86). The relationship of the VPD to the ECDC is communicated by the presence of the ECDC logo in the top two-thirds of the profile picture, while the inclusion of the EU flag in the ECDC logo emphasizes its relationship to the EU.

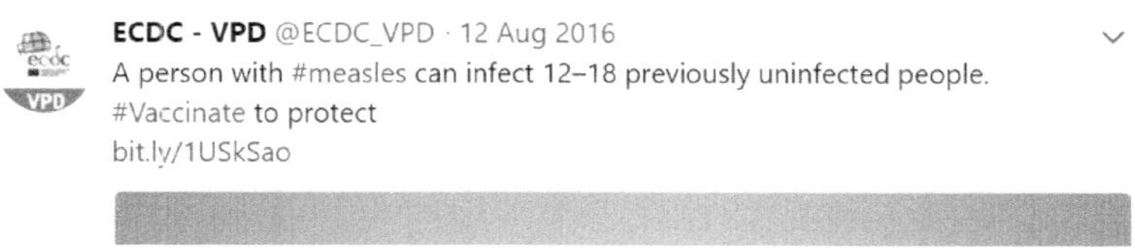

Figure 1. Appeals to credibility in an @ECDC-VPD tweet (ECDC 2005–2019)

We see a similar expression of authority when we look at the cover of a typical ECDC report. Figure 2 shows the masthead of the title page of an ECDC surveillance report. The ECDC logo is displayed in the top right-hand corner of the page and the title of the report is presented in a bold, white, plain font against a green background, with the words "surveillance report" written in capitals above. Once again, the font and typeface suggest authority.

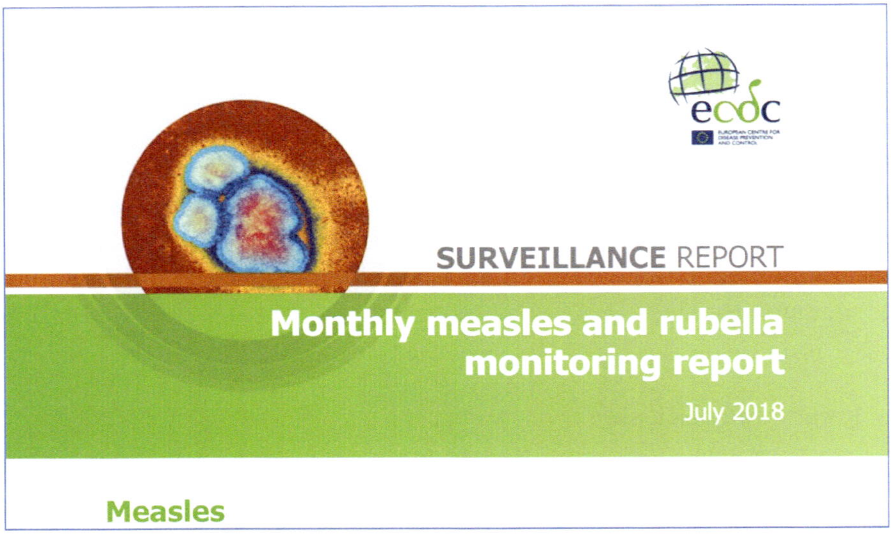

Figure 2. Masthead of title of ECDC surveillance report (ECDC 2005–2019)

The credibility of the information presented in technical reports is reinforced by the expression of the professional status and affiliation of the authors and other collaborators. One such report gives the author as Dr Cornelia Betsch, University of Erfurt, Germany. A similar strategy is used in the @ECDC_VPD tweets, only in this case it is the status of the sources of retweets which enhance the expression of authority. Just over 51% of messages on the @ECDC_VPD account are retweets and most of them originate from other health agencies: 38% from EU agencies, such as the ECDC itself and EU Health; 17% from WHO-Europe; 4% from national health agencies. A further 14% of retweets originate from academic institutions, for example, the Robert Koch Institute, or from academic journals, for example, the *New England Journal of Medicine*.

5.2 Organization and argument

The target audience at which the ECDC reports are aimed comprises professionals with a high degree of expertise. Unlike the research article, the main claims in ECDC reports are presented not in the context of a particular research tradition but in the context of existing EU health policy aims and priorities. For example, the ECDC technical report titled, "Measles and Rubella Elimination: Communicating the Importance of Vaccination," published in 2014, begins with a statement of aims, in which the goals of the ECDC and ECDC-VPD are foregrounded. It states, "The goal of this project is to support EU Member States in their fight against measles and rubella."

Turning attention to the @ECDC_VPD tweets, one of their key functions is to serve as a vehicle for the dissemination ECDC surveillance and technical reports. As a result, 81.4% of the @ECDC_VPD-authored tweets contain a link to an ECDC webpage. The majority of these webpages (80.2%) contain press releases or summaries of the findings of the latest ECDC reports and 72.7% of these contain links to the reports themselves. A mere 1.6% of @ECDC_VPD-authored tweets link to research articles published in scientific journals. If we examine the @ECDC_VPD tweets which link to, and thus draw directly on, ECDC reports, we find that their structure is similar in some respects to the science blog: the newsworthy claim (or an implicit claim to newsworthiness) is placed at the front of the text. The simplest structure is *nominal group/clause + hyperlink*. The nominal group or clause functions merely to point forward to the hyperlink, signal its presence and highlight the newsworthiness of the linked text or the main claim contained in the text, for example:

(1) [*Introductory nominal group*] INVS measles data update [*Hyperlink*] bit.ly/1FsBhWp.

Often, though, the introductory clause encodes newsworthiness by highlighting one of the key claims from the source text. In Example 2, the newsworthy claim is that the number of measles cases have fallen in comparison with the previous year.

(2) [*Introductory clause*] #Measles cases (4735) in 30 #EU countries significantly less than the preceding 12-month period (10 006) [*Hyperlink*] http://bit.ly/1rZvLVE RT.

The majority of tweets are variations of this basic structure, as Examples 3 and 4 show. In Example 3, two clauses precede the hyperlink: the first expresses the main claim while the second describes the type of material in the linked text and attributes the source. In Example 4, the main claim is expressed in two clauses which form a General-Particular pattern: the ideational content of the second clause is an elaboration of the ideational content of the first. The hyperlink is also prefaced with a directive.

(3) [*Main claim*] Studying human behaviour is key to understanding vaccine hesitancy: [*Description of source and attribution*] Interview with anthropologist Heidi Larson [*Hyperlink*] http://bit.ly/MDunKj.

(4) [*Main claim: General*] In 2014, 7 EU countries reported #measles outbreaks. [*Main claim: Particular*] New outbreak now reported in Slovenia. [*Directive*] See latest #CDTR: [*Hyperlink*] http://bit.ly/1AaOYJb.

What is interesting is that, over time, the structure of many tweets has become more elaborate and an increasing number follow an SPSE pattern. This tendency has increased in particular since November 2017, when Twitter raised its character limit to 280. Example 5 dates from 16 July 2018.

(5) [*Situation + Problem: General*] #Measles continues to circulate in the EU/EEA, [*Problem: Particular*] with new outbreaks reported. [*Problem: Particular*] To date, 31 deaths have been reported across the region in 2018. [*Solution + Evaluation*] #Vaccination is the most effective way to stop spread of measles. [*Nominal group introducing hyperlink*] Latest @ECDC_EU data: [*Hyperlink*] http://bit.ly/2uppXPp.

In Example 5, the first clause describes a situation in which a problem is implicit: *#Measles continues to circulate in the EU/EEA*. The adjunct *with new outbreaks reported* provides an elaboration of the problem and stands in relation to the first clause as the "Particular" meaning element in a General-Particular pattern. The second clause, *To date, 31 deaths have been reported across the region in 2018*, also provides Particular information. The third clause, *#Vaccination is the most effective way to stop spread of measles*, presents vaccination as a solution to the problem, while the phrase *the most effective way* expresses a positive evaluation of the solution. The final clause complex, *Latest @ECDC_EU data: http://bit.ly/2uppXPp*, consists of the hyperlink preceded by an introductory clause. What is also noticeable in this example is the inclusion of explicit markers of cohesion to help the reader make sense of the text, such as the preposition *with* in the prepositional phrase *with new cases reported* and the prepositional phrase *to date* at the start of the second clause.

Graphic images, typically maps, graphs, bar charts and tables, are frequently used in ECDC surveillance and technical or expert opinion reports. Since the target audience has a high degree of expertise, graphic images are used to provide supporting evidence. The tweets which appear on the @ECDC_VPD account have shown a growing tendency over time to include images. While, in the first six months of 2014, a mere 18% of tweets and retweets contained an image, in the corresponding period in 2018, the proportion had risen to 91%. The extent to which images function to provide supporting evidence for a claim, as in research articles (Miller 1998), or to entice the reader and simplify concepts, as in popularizations (ibid.), varies according to who the presumed audience is. Tweets which link, via the ECDC website, to a surveillance or technical report are aimed largely at the same expert audience for whom the reports are written. Images in such tweets are reproductions of images contained in the report and function to support a newsworthy claim selected from the report. The example in Figure 3 illustrates this tendency.

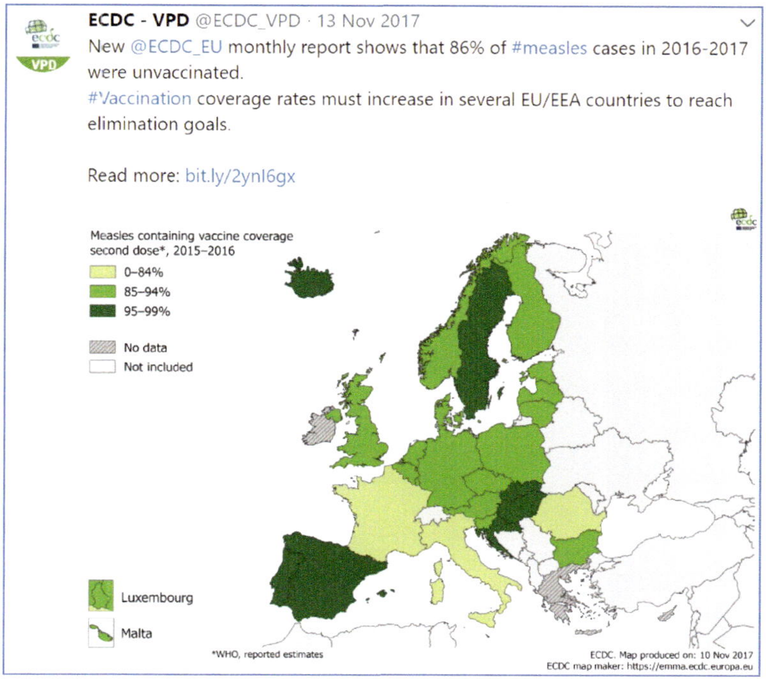

Figure 3. Tweet with graphic image and link to monthly surveillance report (ECDC 2005–2019)

The newsworthy claim reproduced in Figure 3 is that 86% of measles cases in the EU in 2016–17 occurred among unvaccinated individuals. The image shows a map of Europe, with different EU/EEA countries depicted in darker or paler shades of green (or dark and light grey where no data is available), according to the rate of uptake of measles vaccine in 2015–16. The image is constructed in the standard fashion, with given information on the left and new information on the right (Kress and van Leeuwen 1996: 187). The key in the top left corner describes the different colours used to represent different percentage rates of coverage with the second dose of measles vaccine: pale green for 0% to 84%; mid-green for 85% to 94%; and dark green for 95% to 100%. The image of the map to the right of the key presents new information. The image fulfils two functions. With its clearly articulated colour scheme, it provides evidence to support the claim, just as it does in the surveillance report in which it originates. At the same time, it attracts the reader's attention.

The tweet shown in Figure 4, on the other hand, is intended for a wider and more diverse audience. The written text is aimed at exhorting individuals to vaccinate. It also contains a hyperlink which connects to an ECDC web page publicizing the launch of European Immunization Week, 2018.

Figure 4. Tweet aimed at non-expert audience (ECDC 2005–2019)

Here, the visual image is designed to entice the reader and simplify the messages, as is the case in popularizations (Miller 1998). It shows a blue cartoon cow against a bright yellow background, with written text in red below. The cartoon cow is depicted with a smile on its face and two syringes, in dark and light pink, in its rear. The cow functions as a visual metaphor for a person who has been immunized with two doses and is thus happy and healthy. As such, it is typical of health promotion literature (Maci 2014). The bright, heavily saturated colours draw attention to the image and suggest truth (Machin 2007: 59). The image is constructed according to a typical "real-ideal" pattern (Kress and van Leeuwen 1996: 193): what is "real" is represented by the written text in the lower half of the image; the cow in the upper half of the image represents the ideal (being immunized). The written text in the image reformulates the information in the body of the tweet.

5.3 Stance and engagement

5.3.1 *The ECDC reports corpus*

The nature of the ECDC reports impacts on the expression of stance in the data. The reports are intended for internal consumption, that is, for use by health professionals and policy makers employed in health institutions in EU member states. So, where research article authors typically hedge a lot of their claims in order to suggest plausible reasoning and avoid "inviting rejection of their arguments"

(Hyland 2010: 123), the ECDC report authors have less need to do so. This is very much the case with the surveillance reports, which serve mainly to furnish data and are not predominantly argumentative texts. As a result, there is less evidence of use of hedging devices in the ECDC Reports corpus than might be expected. This is illustrated, firstly, by the large number of epistemic modal verbs and adverbs in the corpus which are negatively key. These are *cannot* (frequency 28; LL −29.49); *might* (86; −66.96); *probably* (17; −78.29); *perhaps* (10; −143.35); *could* (181; −208.59); *would* (213; −494.60). The only expressions which signal possibility or probability and which are positively key are *risk* (648; 2,461.27) and *likelihood* (33; 84.94), their keyness in the corpus being attributable to the fact that assessments of probability and risk are a central concern in epidemiology.

Examination of the most frequent n-grams used to express degrees of certainty bears out the conclusion that the ECDC report authors tend to express claims with a relatively high degree of certainty. Table 1 shows, in the left-hand column, the 20 most frequently occurring two- to three-word n-grams containing an unmodalized verb or auxiliary verb or the modal *will*. In the right-hand column, it shows the 20 most frequent two- to three-word n-grams containing words or phrases which indicate possibility, uncertainty or tentativeness.

Table 1. The twenty most frequent two- to three-word n-grams in the ECDC Reports corpus used to express, respectively, certainty and possibility

	Certainty			Possibility	
	N-gram	**Freq.**		**N-gram**	**Freq.**
1	it is (a)	205	1	(the/at/a) risk of	174
2	did not	145	2	can be	150
3	there is (no/a)	141	3	could be	45
4	were reported	133	4	increased risk (of)	42
5	do not	104	5	would be	58
6	it was	86	6	(a/at) high risk (of)	37
7	there was	74	7	likely to	37
8	member states reported	66	8	(the) risks of	30
9	are not	61	9	could not	19
10	will be	50	10	more likely (to)	17
11	were not	48	11	can/may lead to	16
12	(BE) caused by	47	12	low risk (of)	15
13	(was/were) found to	46	13	suggest that	13
14	does not	45	14	that may	13
15	they are	45	15	might be	12

(continued)

Table 1. *(continued)*

	Certainty			Possibility	
	N-gram	Freq.		N-gram	Freq.
16	(had/has) led to	44	16	appear to	11
17	stated that	43	17	can cause	11
18	was found	39	18	cannot be	11
19	was not	38	19	which could	11
20	have been reported	37	20	appears to	10

As can be seen in Table 1, not only are expressions indicating possibility or uncertainty less frequent than those indicating certainty, but many of the expressions of possibility involve the word *risk*. Expressions such as *at/the/a risk of* and *increased/high risk of* function more to express an assessment of the likelihood of an undesirable outcome than as a way of hedging a claim, as the following corpus example illustrates.

(6) At least one million children are currently *at risk of contracting diphtheria* as a result of the unavailability of vaccines and medicines in Yemen.

Admittedly, hedging devices may be found in the co-textual environment of strings such as *it is*, *there is*, *it was*, and so on. However, concordance evidence shows little evidence of this. Statements involving *it is* are hedged in a mere 19 instances out of 205, for example *it is (also) possible* (8 occurrences), *it is usually (given)* (5), while there are only two occurrences of hedging out of the 86 instances of *it was* and one out of the 141 citations of *there is*. What is frequent in the data are unhedged claims about the world (see Example 7) or, more usually, about scientific findings (see Examples 8 and 9).

(7) *It is a common misperception* that measles is a harmless disease.

(8) Interventions that focus on allowing parents to ask questions [...] were well received but *did not have* any effect on patient hesitancy.

(9) *There is no evidence* to date that rotavirus vaccination programmes are driving the emergence of vaccine escape strains.

Where hedging does occur in the corpus, it is more likely to be found in the technical than the surveillance reports. This is because the technical reports contain a greater element of discussion and argumentation and so display more features typical of research articles than the surveillance reports, as Example 10 illustrates.

(10) This *could be explained* by secular trends, population immunity, host and environmental factors, and the emergence of new clones.

Relatively few items occur in the data which function explicitly as boosters, but, as we would expect of scientific reports, we find comparison and intensification through the use of quantification. These phenomena are signalled by the presence of phrases such as *the most* (121 occurrences), *more than* (64 occurrences), *the highest* (45) or *less than* (29), for example:

(11) *The most common determinant* of non-vaccination was the belief that vaccines are unsafe.

(12) ... *more than 28,000 doses* were administered.

As for attitude markers, vaccines are often evaluated in terms of their effectiveness, while policies are evaluated, not just in terms of how successful they are, but in terms of perceived value for money. Hence, we find *effectiveness* (217), *effective* (137), *cost effectiveness* (68), *vaccine effectiveness* (41), and *cost effective* (36). Finally, there are a small number of n-grams which function as modal-like expressions (Hunston 2011), that is, attitude markers which also express deontic meaning. These are:

> *the importance of* (72), *it is important (to)* (74), *is essential to* (20), *is important for* (15), *only effective (way)* (13), and *be useful (for/to)* (9).

Examples of use are:

(13) UNICEF stresses *the importance of* monitoring hesitant populations and websites continuously.

(14) Vaccination is *the only effective way to* protect against measles.

Expressions such as these are modal-like because they encode an evaluation of an object or state of affairs but at the same time recommend a particular course of action.

If we turn now to more explicit markers of engagement (see Table 2), we find that there are few instances of reader pronouns, questions or imperatives. The citations of *do you think (that)* are examples of questions asked in patient questionnaires, while *check your immunity* is an example of advice given to patients. Directives are expressed using indirect forms, either by using a passive construction involving a modal verb, such as *should be*, *must be*, (BE) *required to*, (BE) *needed to*, and so on, or through expressions containing nouns, verbs or adjectives which express necessity or obligation, such as *need to/for*, *is recommended* (*by*), *necessary to*, *ensure that*.

Table 2. Two- to three-word n-grams in the ECDC reports corpus, with a frequency of five or more, which signal engagement

	N-Gram	Freq.		N-Gram	Freq.
1	should be	186	9	do you think (that)	24
2	(the) need to	68	10	necessary to	21
3	(the) need for	60	11	should receive	21
4	to ensure (that)	44	12	should avoid	13
5	must be	38	13	women should	11
6	is recommended (by)	25	14	recommends that	8
7	required to	25	15	check your immunity	5
8	needed to	25			

The preference for using passive forms to convey deontic meaning, or to use forms involving nominalization, may result from the fact that the ECDC reports, as scientific texts, display many features typical of scientific research articles.

5.3.2 *The @ECDC_VPD corpus*

Since many of the tweets in the @ECDC_VPD corpus draw on, and thus refer to, specific ECDC reports, the @ECDC_VPD and ECDC Reports corpora share many of the same features. However, on account of the differing communicative purposes to which the reports and the tweets are put and owing to the space constraints of Twitter, there are crucial differences between the two. Firstly, where, in the ECDC Reports corpus, there were several occurrences of expressions of certainty but somewhat fewer of uncertainty or possibility, there is a similar pattern in the @ECDC_VPD corpus but the number of different expressions used in each category is far more limited (see Table 3). Furthermore, there are a mere thirty occurrences of expressions indicating possibility.

Table 3. Two- to three-word n-grams in the @ECDC_VPD corpus, with a frequency of five or more, which express certainty or possibility

Category	N-gram with frequency
possibility	can be (7); risk of (7); could be (6); measles can (5); we can (5)
certainty	vaccination is (22); will be (22); is not (17); is now (9); (cases) were reported (in) (9); is still (6); measles affects all (6); measles is more (6), vaccine is (6), continues to (5); is most (5); is the most (5)

Of course, the relatively restricted variety of expressions may be due to the fact that the @ECDC_VPD corpus is far smaller than the ECDC Reports corpus. The preference for un-modalized expressions, though, is likely to be due to the need for the @ECDC_VPD authors to communicate a clear, unambiguous and attention-grabbing message in a space-constrained context, as is illustrated in the following examples.

(15) *#Vaccination is not just for kids.* Adults need protection too.

(16) *#measles affects all age groups* #immunizeurope.

The relative frequency of un-modalized assertions appears far higher if we include the many non-finite clauses which are used in the corpus data, no doubt as a response to the space constraints. In Example 17, not only is a non-finite clause employed but abbreviations (*btw* and *sept*) are used.

(17) *2266 measles cases reported in* EU/EEA btw sept 2015 and august 2016.

The second difference between the @ECDC_VPD and ECDC Reports corpora is that, in the former, we can find a greater number of explicit markers of attitude and engagement among the keywords (see Table 4).

Table 4. Positive keywords in the @ECDC_VPD corpus which function as markers of attitude and engagement

Category	Keyword with frequency and log-likelihood
attitude	new (138; 155.78); #vaccineswork (82; 1327.35); latest (75; 398.69); #immunizeeurope (66; 1068.35); effective (35; 107.26); effectiveness (20; 992.3); safe (13; 26.20)
engagement	read (66; 181.61); join (34; 121.71); check (28; 91.26); submit (12; 59.73); watch (20; 44.32); follow (19; 38.19).

As we saw with the ECDC Reports corpus, expressions of attitudinal meaning are often used in contexts in which deontic meaning is expressed. The words *new* and *latest*, for example, tend to occur in contexts in which the novelty of a piece of information is highlighted. The n-grams in which they participate are:

> *latest ecdc (eu)* (17); *new ecdc (eu)* (17); *the latest* (13); *a new* (9); *the new* (9); *bmj latest* (7); *new report* (7); *new data* (6); *new vaccine* (5); *new vaccination* (5); *read latest* (5).

Concordance evidence reveals that *new* is often found at the start of a tweet which includes a link to a report. It functions to emphasize the newsworthiness of the main claim in the report. In most cases, the final clause in the tweet contains an imperative, typically *read* or *read more*, as can be seen in Example 18.

(18) *New @ECDC_EU #measles data:* outbreaks still ongoing in Europe, 48 deaths since January 2016. [...] *Read more*: https://t.co/l7NG3oD1vJ https://t.co/mfEGPOdcDJ.

The word *latest* is used to highlight the novelty not of a claim but of a particular text. *Latest* occurs either in clauses which start with an imperative, again this is the verb *read* (see Example 19), or in nominal groups which introduce a hyperlink (see Example 20). In the latter case, the nominal group in which *latest* participates functions as a directive to click on the hyperlink and read the material.

(19) *Read latest report* from ECDC: http://t.co/O3nPUbi2xs.

(20) *Latest ECDC* #STEC, #EHEC *update* (22 June, 11:00) http://bit.ly/lRH0ul#ECDC.

As the data in Table 4 showed, a number of explicit directives are used relatively frequently in the @ECDC_VPD corpus. *Watch* and *follow* are used in the same way as *read*, that is, they instruct the user to watch a certain video or follow a particular Twitter account. The fact that imperatives are used in these instances may be due to the fact that this is a typical feature of Twitter interaction (Knight 2015: 29) or it may be because of the low ranking of imposition of these particular face-threatening acts (Brown and Levinson 1987). Where tweets are aimed at the vaccine-hesitant, explicit directives are sometimes used, for example, *check your vaccination status* (6) or *get vaccinated* (5), but, more often, encouragement to vaccinate is expressed using the sorts of modal-like expressions which were observed in the ECDC Reports corpus, namely *most effective* (11), *needed to* (8), *the importance of* (8) and *best protection* (5), as Examples 21 and 22 illustrate.

(21) Significant #measles outbreaks ongoing in EU – #vaccination is *most effective way to* stop spread of the disease.

(22) 2 doses of vaccine *needed to ensure full protection.*

One final area of similarity lies in the use of intensification expressed using quantification. Usually, in the @ECDC_VPD tweets, precise numbers are not given but an approximation preceded by an expression such as *more than* or *over* (see Example 23). The choice of an approximate figure modified by an intensifying expression such as *more than* or *over* emphasizes the sheer scale of the problem in a way that giving a precise figure might not.

(23) *More than 9500 #measles* cases were reported by 30 EU countries fr 04/2103 to 03/2014.

The analysis of stance and engagement strategies in the @ECDC_VPD corpus undertaken here has revealed that a limited range of expressions of stance markers are used with any degree of frequency. For example, the words *new* and *latest* are used to the exclusion of other possible choices to highlight the newsworthiness of claims or of hyperlinked texts. As for engagement, indirect strategies are preferred. Explicit directives are used in instances where the threat to face is low, but exhortations to the public to vaccinate are typically expressed using an off-record strategy (Brown and Levinson 1987). The reason for this may be that asking someone to vaccinate implies asking them to accept an invasive procedure. Furthermore, when it comes to childhood vaccinations, a parent may feel conflicted by a perception of competing risks: the risk of your child being harmed by the vaccine as opposed to the potential risks posed by leaving your child unprotected against harmful diseases.

6. Conclusion

The analysis has shown that the @ECDC_VPD authors tend to draw on the source texts by selecting a newsworthy claim and foregrounding it by placing it at the top of the written textual element of the tweet. Interestingly, the macro-structural properties of @ECDC_VPD tweets have increased in complexity over time. The simple strategy of embedding a hyperlink in the tweet preceded by an introductory nominal group or clause has given way to multi-clausal structures. A clause expressing a generalized claim is now more likely to be followed by a clause expressing a more detailed claim. Since Twitter expanded its character limit, writers increasingly construct tweets according to an SPSE pattern and often include within that structure clauses paired in a General-Particular relation. As a result, more complex meanings are communicated. The reader is also guided through the text since clause relations are made explicit and familiar argument structures are employed.

The fact that the primary audience for the @ECDC_VPD tweets consists of health professionals and policy makers accounts for the fact that, in the majority of tweets containing visual elements, images tend to be drawn from ECDC reports and provide evidence to support a claim. However, it could be argued that, since such tweets provide links to entire texts, the image is largely redundant. If that is so, it becomes clear that the main function of images in these tweets is to attract the reader's attention. This is certainly the case when it comes to images in tweets aimed at the general public. These function much like images in public health leaflets, in that they simplify and explain (Maci 2014).

Although there was found to be a slightly wider range of attitude markers in the @ECDC_VPD corpus data than in the ECDC Reports corpus, the difference does not lend weight to the suggestion that science discourse on Twitter is necessarily marked by a relatively higher occurrence of attitude markers than is found in other genres (Veltri and Atanasova 2015; Büchi 2016), although this might be due to the fact that the @ECDC_VPD Twitter account represents an international health agency and the writers would be aware of the need to uphold the credibility of the organization. The relatively low occurrence of modals in @ECDC_VPD corpus reflects tendencies that have been noted in Twitter discourse elsewhere (Knight 2015). The core activity of the ECDC-VPD concerns the dissemination of disease surveillance data. It is therefore unsurprising that both the @ECDC_VPD and ECDC Reports corpora contain several instances of quantification. In the @ ECDC_VPD corpus, however, the drive for newsworthiness means that numerical expressions are frequently modified by an intensifying expression.

The fact that a greater range of engagement strategies was observed in the @ECDC_VPD corpus than in ECDC Reports is no doubt due to the interactive nature of Twitter. It was found that in the @ECDC_VPD tweets directives are often used, especially when the ranking of imposition is low. These findings might be explained by Knight's (2015) observation that the relative anonymity of Twitter, together with space constraints, makes the use of indirect politeness strategies less likely. Indirect strategies, however, were found to be preferred in @ ECDC_VPD data in contexts in which members of the public are encouraged to vaccinate. Presumably, asking someone to get themselves or their child vaccinated constitutes a greater imposition than asking them to click on a hyperlink and read a text. Furthermore, the members of the public the ECDC-VPD hopes to reach are the vaccine-hesitant. It is probably on account of these factors that modal-like expressions are preferred over directives. In terms of politeness, we can class the use of a modal-like expression to realize deontic meaning as a form of off-record strategy. What is interesting about the use of modal-like expressions in the @ ECDC_VPD data is that they manage to conflate affective and deontic meaning and so represent a useful strategy for managing space constraints.

By focusing on the relationship between the Twitter messages of an international health agency and the scientific reports on which they draw, this chapter has shed light on an area which has received little attention in popularization studies. A limitation of the study is that it does not consider the wider textual networks in which the @ECDC_VPD tweets participate. This would be a potentially fruitful avenue for future research.

References

Anthony, Laurence. 2019. AntConc (Version 3.5.8) [Computer Software]. Tokyo, Japan: Waseda University. Available from http://www.laurenceanthony.net/software/antconc/

Anthony, Laurence, and Claire Hardaker. 2017. FireAnt (Version 1.1.4) [Computer Software]. Tokyo: Waseda University. Available from http://www.laurenceanthony.net/software

Brown, Penelope, and Stephen Levinson. 1987. *Politeness*. Cambridge: Cambridge University Press. https://doi.org/10.1017/CBO9780511813085

Büchi, Moritz. 2016. "Microblogging as an Extension of Science Reporting." *Public Understanding of Science* 26 (8): 953–968. https://doi.org/10.1177/0963662516657794

Caliendo, Giuditta. 2014. "The Popularization of Science in Web-based Genres." In *The Language of Popularization: Theoretical and Descriptive Models*, ed. by Giancarmine Bongo, and Giuditta Caliendo, 101–132. Bern: Peter Lang.

Calsamiglia, Helena, and Teun van Dijk. 2004. "Popularization Discourse and Knowledge about the Genome." *Discourse and Society* 15 (4): 369–389. https://doi.org/10.1177/0957926504043705

ECDC. (2005–2019). Copyright and Limited Reproduction Notices. Retrieved 7 June 2019. https://ecdc.europa.eu/en/copyright

Fahnestock, Jeanne. 1986. "Accommodating Science: The Rhetorical Life of Scientific Facts." *Written Communication* 3 (3): 275–296. https://doi.org/10.1177/0741088386003003001

Hilgartner, Stephen. 1990. "The Dominant View of Popularization: Conceptual Problems, Political Uses." *Social Studies of Science* 20 (3): 519–539. https://doi.org/10.1177/030631290020003006

Hoey, Michael. 1983. *On the Surface of Discourse*. London: Allen and Unwin.

Hunston, Susan. 2011. *Corpus Approaches to Evaluation: Phraseology and Evaluative Language*. London: Routledge.

Hyland, Ken. 2005. "Stance and Engagement: A Model of Interaction in Academic Discourse." *Discourse Studies* 7 (2): 173–192. https://doi.org/10.1177/1461445605050365

Hyland, Ken. 2010. "Constructing Proximity: Relating to Readers in Popular and Professional Science." *English for Academic Purposes* 9 (2): 116–127. https://doi.org/10.1016/j.jeap.2010.02.003

Knight, Dawn. 2015. "e-Language: Communication in the Digital Age." In *Corpora and Discourse Studies: Integrating Discourse and Corpora*, ed. by Paul Baker, and Tony McEnery, 20–40. London: Palgrave Macmillan. https://doi.org/10.1057/9781137431738_2

Kramarz, Piotr, Pierluigi Lopalco, Emma Huitric, and Lucia Pastore-Celentano. 2014. "Vaccine-preventable Diseases: The Role of the European Centre for Disease Prevention and Control." *Clinical Microbiology and Infection* 20 (s5): 2–6. https://doi.org/10.1111/1469-0691.12430

Kress, Gunther, and Theo van Leeuwen. 1996. *Reading Images: The Grammar of Visual Design*. London: Routledge.

Luzón, María-José. 2013. "Public Communication of Science in Blogs: Recontextualizing Scientific Discourse for a Diversified Audience." *Written Communication* 30 (4): 428–457. https://doi.org/10.1177/0741088313493610

Machin, David. 2007. *Introduction to Multimodal Analysis*. London: Hodder Arnold.

Maci, Stefania. 2014. "Institutional Popularization of Medical Knowledge: The Case of Pandemic Influenza A (H1N1)." In *The Language of Popularization: Theoretical and Descriptive Models*, ed. by Giancarmine Bongo, and Giuditta Caliendo, 165–189. Bern: Peter Lang.

Miller, Thomas. 1998. "Visual Persuasion: A Comparison of Visuals in Academic Texts and the Popular Press." *English for Specific Purposes* 17 (1): 29–46. https://doi.org/10.1016/S0889-4906(97)00029-X

Puschmann, Cornelius. 2014. "(Micro)blogging Science? Notes on Potentials and Constraints of New Forms of Scholarly Communication." In *Opening Science*, ed. by Sascha Friesike, and Sönke Bartling, 89–106. New York: Springer. https://doi.org/10.1007/978-3-319-00026-8_6

Scotto di Carlo, Giuseppina. 2014. "The Role of Proximity in Online Popularizations: The Case of TED Talks." *Discourse Studies* 16 (5): 591–606. https://doi.org/10.1177/1461445614538565

Shan, Liran, Áine Regan, Aoife De Brún, Julie Barnett, Maarten C. A. van der Sanden, Patrick Wall, and Áine McConnon. 2013. "Food Crisis Coverage by Social and Traditional Media: A Case Study of the 2008 Irish Dioxin Crisis." *Public Understanding of Science* 23 (8): 911–928. https://doi.org/10.1177/0963662512472315

Titak, Ashley, and Audrey Roberson. 2013. "Dimensions of Web Registers: An Exploratory Multi-dimensional Comparison." *Corpora* 8 (2): 235–260. https://doi.org/10.3366/cor.2013.0042

Veltri, Giuseppe A., and Dimitrinka Atanasova. 2015. "Climate Change on Twitter: Content, Media Ecology and Information Sharing Behaviour." *Public Understanding of Science* 26 (6): 721–737. https://doi.org/10.1177/0963662515613702

CHAPTER 10

The representation of science and technology in genres of Vatican discourse

Pope Francis's encyclical *Laudato Si'* as a case study

Graham Smart and Matthew Falconer
Carleton University

The study presented in this chapter focuses on the discursive representations of science and technology embedded in *Laudato Si'*, an instance of a traditional print genre in the process of becoming digitalized. The study describes the epistemic discursive activity organized by two genre sets, with *Laudato Si'* occupying a place in both. The first genre set, which we have labelled the "science/technology genre set," has functioned for almost a century as a discursive vehicle supporting an intertextual chain of uptake and recontextualization used in developing an evolving body of institutional knowledge and collective argumentation regarding science and technology, knowledge and argumentation that eventually emerged in *Laudato Si'*. The second digital genre set, which we refer to as the "launch-day genre set" was employed to rapidly disseminate *Laudato Si'* around the world. As well, we characterize *Laudato Si'* as an instance of a "hybrid genre," a genre combining certain affordances of digital communication with valued residual features of an antecedent genre and its traditional print text.

Keywords: papal encyclical, digital genre, genre set, representation, science and technology, recontextualization

1. Introduction

On June 18, 2015, Pope Francis's encyclical letter, *Laudato Si', on Care for our Common Home'* (hereafter *Laudato Si'*) was published digitally on the Internet by the Vatican Press Office.[1] With the Paris Climate Change Conference fast

1. While a number of meanings have been ascribed to the term "digital," in this chapter we adopt a definition from the online *Longman Dictionary of Contemporary English*: "Using a system in which information is recorded or sent out electronically in the form of numbers, usually ones and zeros."

https://doi.org/10.1075/pbns.308.10sma
© 2019 John Benjamins Publishing Company

approaching in November 2015, this carefully chosen date of digital publication was by design a kairotic moment for Francis and his encyclical, and a key part of the Pope's effort to reach a diverse worldwide audience with his impassioned plea for an "integral ecology" – a holistic vision of ecology interconnecting the Earth's natural environment and various aspects of human society.

Pope Francis's 183-page *Laudato Si'*, an instance of a traditional print genre in the process of becoming digitalized, serves as a centerpiece for the study presented in this chapter. More specifically, the study focuses on discursive representations of science and technology (henceforth "science/technology") embedded in *Laudato Si'*, viewing these representations as the product of a historical dialogue featuring, on one side, a succession of eight popes, beginning with Pope Pius XI in 1917 and continuing through to Pope Francis, and on the other side, an international cross-disciplinary group of natural scientists belonging to the Pontifical Academy of Sciences (hereafter "Academy"), the Vatican's in-house scientific advisory body.[2]

Our purpose in the study is two-fold. First, we trace a genealogy – that is, we look back in time to identify the historical roots – of the multi-faceted discursive representation of science/technology contained in the encyclical *Laudato Si'*. Here we consider how this representation of science/technology had been evolving for almost a century as part of the Vatican's side of the dialogue mentioned above, to eventually emerge as a key aspect of *Laudato Si'*. In looking at how this representation of science/technology has evolved over time, we see how it has grown in ideational complexity over the decades.

In tracing this genealogy of the representation of science/technology exhibited in *Laudato Si'*, we describe a body of discourse that we call the "science/technology" genre set, which for over a century has served as a discursive vehicle for the long-running dialogue between the post-1916 succession of popes and generations of Academy scientists, a dialogue reflecting their shared concern for keeping abreast of advances in science/technology and for reflecting on its appropriate role in contemporary society. We also show how the evolving knowledge and collective argumentation inherent in representations of science/technology that have been produced through one part of the genre set mentioned above – the *papal address* delivered biennially to the Academy's scientists – can be viewed as a product of a knowledge-generating intertextual chain of discursive uptake and recontextualization, begun in 1917 with the founding of the Academy and continuing through to and beyond the digital release of *Laudato Si'* in June 2015. Here we see an occurrence of the necessary interdependence among genres constituting a genre

2. In our research, we found that in references to "science" and "technology" in Vatican discourse, the line between the two was often blurred. For this reason and for the gain in conciseness, we use the term "science/technology" throughout the chapter.

set that has been designed to create and communicate specialized knowledge and collective argumentation (Smart 2011) – that is, argumentation developed and communicated by a group of social actors functioning together discursively.

Second, we look forward in time from the digital release of *Laudato Si'*, enacted by another body of discourse that we refer to as the "launch-day" genre set, to examine the numerous and diverse digital textual responses, communicated in a number of genres, that were evoked worldwide by the appearance of the Pope's *Encyclical* and posted by their authors in the public space of the Internet. In this part of the chapter we focus on those particular textual responses whose authors have taken up and recontextualized different aspects of *Laudato Si'*s representation of science/technology for the authors' own purposes of supporting or contesting the Pope's impassioned global call-to-action for an "integral ecology."

Four research questions have guided our study:

1. How can the discursive representation of science/technology inherent in *Laudato Si'* be characterized, and what are the historical roots of this discursive representation?
2. What recurrent patterns of uptake and recontextualization can be found in the body of discourse – viewed as two interconnected genre sets – associated with *Laudato Si'*?
3. What new affordances did the digital version of the papal encyclical genre offer the Pope and the Vatican beyond those affordances of the traditional print version of the genre?
4. What does it mean to call *Laudato Si'* an instance of a hybrid genre?

From here, we first provide some background regarding the genre of the papal encyclical and the history and work of the Pontifical Academy of Sciences. We then discuss the theoretical perspectives that inform our study, describe our research method, present our findings, and conclude with some implications of this research.

2. Background

2.1 The genre of the papal encyclical

Used in ancient Greek and Roman times to refer to any geographically circulating letter (*The Catholic Encyclopedia*, 2017), the term "papal encyclical" (hereafter "encyclical") later took on a more specific meaning within the tradition and practices of the Catholic Church (the Church hereafter). As a genre, the encyclical assumed its modern function and form in 1740 when Pope Benedict XIV sent an encyclical titled, *Epistola Encyclica et Commonitoria* (*On the Duties of Bishops*), to archbishops, bishops, and other senior clerics in the various geographical regions of the Church. Since then, encyclicals have been widely used by popes, and popes only, to convey a message to the larger membership of the Church, a message usually regarding a significant matter of doctrine or moral conduct, while also serving as a teaching resource within the various congregations of the Church. The encyclical has been described as "the highest and most comprehensive level of teaching in the Catholic Church" (Grim 2015). Since 1740 some 250 encyclicals have been issued by fifteen popes.

Traditionally, when an archbishop received an encyclical from the Pope, he would have the print document copied and sent out to the bishops in his region, who would arrange to have the contents duplicated and communicated to the senior clerics of local churches and their congregations. This activity required much time and effort, so that encyclicals were limited in frequency to messages that popes considered of great import to the Church as a whole. More recently, in a trend initiated by Pope Benedict XVI and extended by Pope Francis, encyclicals have been posted to the Vatican website in various languages and also distributed through other digital media, with a consequent broadening and diversification of the potential audience. Historically, the intended audience for encyclicals has been the larger membership of the Church – archbishops, bishops, other senior clerics, and parishioners – at least until the release of *Laudato Si'*, when the intended audience was much wider than it had been traditionally, as indicated by Pope Francis at the beginning of the *Encyclical*: "Now, faced as we are with global environmental deterioration, I wish to address every person living on this planet. … In this *Encyclical*, I would like to enter into dialogue with all people about our common home" (*Laudato Si'*, 4).

2.2 *Laudato Si'* and the plea for an integral ecology

The concept of an "integral ecology" espoused by Pope Francis in *Laudato Si'* is a much extended understanding of the term "ecology," as the word is commonly used in referring to the symbiotic relationship between living organisms and their natural environment. Francis's notion of an "integral ecology" is intellectually complex, a gestalt as it were, comprising three inter-related dimensions: first, a comprehensive holistic vision of reality that recognizes a web of connections among the natural environment and the scientific, cultural, economic, political, and moral dimensions of human society; second, drawing on this multi-faceted view of reality, an analysis of the multiple deep-rooted and inter-related existential crises that simultaneously threaten the contemporary world – crises of severe economic disparity, cultural disruption, degradation of the natural environment, rampant social injustice, intergenerational inequity, and failing national states – all of which are exacerbated by the massive and pressing threat of climate change; and third, a proposed approach for confronting these crises, an approach involving a radically new way for humans to conceive of reality and live in the world.

For our purposes in this chapter, the most significant aspect of the Pope's integral ecology is the representation of science/technology that forms part of it. In the Pope's vision of an integral ecology, science/technology are both a cause of significant damage to human society and the natural environment and a potential force for mitigation and recovery. According to this representation, science/technology are intimately connected to other aspects of reality mentioned above – scientific, cultural, economic, political, and moral – both as a primary part of the problem and as a possible part of the solution.

2.3 Pontifical academy of sciences

The Pontifical Academy of Sciences was established by the Vatican in 1917, though its historical antecedents go back to 1604, when the Roman Prince Federico Cesi founded the Accademia dei Lincei (the Academy of Lynxes), to provide a forum for reports of scientific experiments, collegial discussion, and mutual support. The Accademia flourished for more than two decades, with Galileo Galilei becoming a leading member, before foundering after Prince Ceci died in 1630.[3] Several failed attempts to re-establish the Accademia ensued before Pope Pius IX revived it in 1847 as the Pontifical Academy of the New Lynxes. In 1917 Pope Pius XI formally resurrected the institution as the Pontifical Academy of Sciences, providing it with statutes regularizing its mandate, membership, and activities.

At present, the Academy's international membership comprises 80–90 scientists from a range of disciplines, and includes men and women of all religions as well as non-believers. Very selective in issuing invitations to join, the Academy has since its inception in 1917 included in its membership over 75 Nobel Prize winners in disciplines such as Physics, Chemistry, Physiology, and Astronomy. According to its first statute, the primary aim of the Academy is to "promote the progress of the mathematical, physical, and natural sciences and the study of related epistemological questions and issues." This goal is accomplished primarily through a program of biennial week-long plenary meetings that bring to Rome all the Academy scientists from around the world to collaborate in addressing a pre-identified scientific topic, as well as through ad hoc Working Groups and Study Weeks of several days duration in which smaller specialized groups of Academy members

3. As a point of interest, the Accademia dei Lincei, established by Prince Ceci in 1604, predates the UK's Royal Society by almost six decades.

come to the Vatican to take up a scientific issue of particular current concern.[4] The Academy also plays an advisory role within the Vatican, as affirmed by Pope John Paul II in his October 1996 address to the Academy: "In celebrating the sixtieth anniversary of the Academy's refoundation, I would like to recall the intentions of my predecessor Pius XI, who wished to surround himself with a select group of scholars, relying on them to inform the Holy See in complete freedom about developments in scientific research, and thereby to assist him in his reflections."

The concluding event in each of the biennial week-long plenary meetings is an audience with the Pope, an event where the President of the Academy, in the company of all the Academy's scientists, summarizes the discussions that have occurred during the previous five days of the plenary meeting, with the Pope then responding in an address to the Academy. This regularly scheduled exchange is intended both to keep the Pope and his colleagues in the Vatican informed about new developments in the natural sciences, and to instruct the Academy's scientists, and by extension the international scientific community, regarding their moral responsibility for directing their work towards the benefit of all humankind. The proceedings of a plenary meeting, which include the responding address from the pope, are published for distribution within and beyond the Vatican in a document known as an *ACTA*, the collection of which is considered to be the Academy's scientific journal.

3. Theoretical perspectives

Conceptually, the research discussed in this chapter draws on three areas of theory: a contemporary understanding of rhetoric, a social conception of genre, and a view of representations that is social constructionist, epistemic, and rhetorical.

3.1 Rhetoric: A contemporary view

For the purposes of our study, we take an expansive view of "rhetoric," one that extends beyond Aristotle's definition of rhetoric as the art of "discovering the best available means of persuasion" in any given communicative situation (*Aristotle*, trans., Cooper, 1960). Seen in this light, rhetoric refers to any effort to achieve a

4. The titles of three recent *ACTA*s provide a sense of the range of topics taken up by the members of the Academy in their biennial plenary meetings: *Science and Sustainability: Impacts of Scientific Knowledge and Technology on Human Society and its Environment*; *Evolving Concepts of Nature*; *Complexity and Analogy in Science: Theoretical, Methodological and Epistemological Aspects.*

particular aim through the use of language (Burke 1969; Bazerman 1988). Some theorists would go further, expanding rhetoric to include, in addition to the use of verbal language, the use of any communicative symbol system to "realize human goals and carry out human activities" (Bazerman 1988: 6). This contemporary perspective on rhetoric carries with it a view of rhetorical strategies that goes beyond Aristotle's appeals of *logos*, *ethos*, and *pathos* and traditional devices such as metaphor, synecdoche, irony, and personification, to include any use of spoken or written language, mathematical equations, graphic images, or other symbols to communicate meaning to an intended audience. For our study, this expanded understanding of rhetoric comes into play in our investigation of the rhetorical and epistemic capacity of verbal representations of science and technology.

3.2 A social conception of genre

Our study of the representation of science/technology in *Laudato Si'*, along with the discursive forerunners of this representation in earlier papal discourse, draws on several aspects of genre theory. We see a "genre" – in this particular case, the genre of the encyclical – as a text-based form of rhetorical action prompted by the demands of a recurrent social situation and exigence (Miller 1984) (see Luzón and Pérez-Llantada this volume). At the same time, we employ the concept of the "genre set" (Devitt 1991; Bazerman 1994) to refer to two or more genres performing different but related rhetorical actions within a common social context, such as, for example, in a school classroom, a corporate head office, a community of social activists, or, in this instance, the Vatican and its in-house science advisory group, the Academy.

3.2.1 *Digital genres*

Miller (2016) suggests that digital genres may evolve out of traditional print genres or, alternatively, may appear suddenly as a new form of discourse, a notion first introduced by Crowston and Williams (2000) and further developed by Herring (2013), with her tri-part classification of "familiar," "reconfigured," and "emergent" genres. In Miller's first case, a genre can often be seen to have grown out of a recognizable "antecedent genre" (Jamieson 1975). In her second case, a new digital genre may appear to have arisen out of nowhere, with no observable antecedent genre in the picture. A third possibility, one involving a situation where a traditional print genre can be seen to be in the process of evolving into a digital genre, occurs with the creation of a "hybrid genre" (Graham and Whelan 2008; Bhatia 2016; Mehlenbacher 2019). For our purposes here we define a hybrid genre, somewhat narrowly, as a partially digitalized genre that, in its transitional state, displays certain features of form and function similar to those of the traditional print genre from which has evolved, while at the same time providing new affordances related to changes in the social context.

3.2.2 *Uptake and recontextualization*

As well, we draw on the concepts of "uptake" and "recontextualization" (Linell 1998; Freadman 2002; Luzón 2013), viewed as a single semiotic process of meaning-making, where a rhetor's uptake of certain meanings from an original text in a given genre and context is a rhetorical act of selection, and where recontextualization is the discursive activity of transforming and repurposing the meanings of the original text in a new instance of discourse – within the same or a different genre – conveying new meanings within a different context (Smart and Falconer forthcoming). In this chapter we use the notion of uptake/recontextualization in three ways: first, in examining the dialogue between the Academy's scientists and the eight popes in office from 1917 to the present, a dialogue embodied in the Academy's *ACTA*s, in the summary report addressed to the pope by the President of the Academy on the discussions of the just-concluded plenary meeting, and in the popes' responding addresses to the Academy; second, in describing the digital launch-day genre set employed to transmit *Laudato Si'* to a worldwide audience in June 2015; and third, in discussing online public responses to *Laudato Si'* following its digital release.

3.3 Representations of science/technology

The nature of representations as discursive phenomena is of obvious significance for our study. Hall (1997) provides us with a basic starting point for thinking about representations, describing them as linguistic or other symbolic elements signifying some material or conceptual aspect of a pre-existing reality. To this basic characterization of representations, Hall goes on to add a social constructionist frame with representations playing a central role, not just in signifying aspects of a pre-existing world, but rather in actually creating a shared social reality and producing related knowledge.

A closely related area of scholarship relevant to our study of representations of science/technology in Vatican discourse is the theory of "social representations" introduced by social psychologist Moscovici (1963). As depicted by Moscovici, a social representation is "a social object [created] by the community for the purpose of behaving and communicating, [an object suffused] with the community's values, ideas and practices" (1963: 251). Potter (1996) offers another perspective on social representations, viewing their use as a discursive practice, with discourse as the environment that social representations inhabit. Billig (1988) concurs with Potter, characterizing social representations as discursive constructions employed for persuasive ends in a world of argumentation where "voices of dispute and controversy are ceaselessly heard" (1988: 9). Applying this discursive perspective on social representations to the public understanding of science, Bauer and Gaskell (1999) claim that the public depends solely on social representations for access to the professional world of scientists and technologists, their expert work, and the specialized knowledge created through this work. In sum, the work described above leaves us with a sense of representations of science/technology as epistemic – that is, infused with knowledge – and serving rhetorical purposes.

4. Method

Between April 2015, two months prior to the digital publication of *Laudato Si'*, and January 2019, primarily using the search engine Google Alerts to gather online textual data, we collected some 1,100 digital texts that were clear responses to *Laudato Si'*, including, in an approximately descending order of occurrence: e-newsletters, e-bulletins, blog posts, news articles, press releases, and tweets. These texts originated with various interest groups, institutions, and individuals, including Church congregations, business associations, NGOs, think-tanks, partisan political groups, and conservative and liberal religious organizations. We also gathered scores of contemporary and historical texts from the websites of the Vatican, including that of the Academy. In analyzing this data, we employed an emergent approach (Hesse-Biber and Leavy 2008): interpreting the data as we gathered it and using the resulting mini-analyses to inform the next phase of the research.

The first step in our analysis was to identify the different "domains of discourse" (Hulme 2009) in Pope Francis's *Laudato Si'*, including, for example, science, politics, the natural environment, economics, religion, and social justice. We then looked closely at all the references to science (13) and technology (33) in the *Encyclical*. Next we did a search for references to science/technology included in a pdf containing some sixty *papal addresses* delivered to the Academy's scientists at the conclusion of their biennial plenary meetings over the decades since 1917. The second step was to scrutinize, manually, the approximately 1,100 responses to *Laudato Si'* published on the Internet, again identifying the "domains of discourse" they contained. Given the purpose of our research, we focused on the content of textual responses in which the central concern was science/technology. Here, using Maarten Hajer's (2006) concept of the "discourse coalition," a group of social actors attracted, within the context a major social debate, to the same set of "story-lines," or arguments, we identified responses to the representations of science/technology in *Laudato Si'* that were supportive and other responses that were highly critical, and at times even hostile.

5. Findings

We begin this section by describing a Vatican genre set that has, historically, functioned as a discursive vehicle for producing and communicating representations of science/technology. This genre set comprises four genres: the *ACTA*, proceedings published following a biennial plenary meeting of the Academy; the *summary report* on the discussions of the week-long plenary meeting, presented to the current pope by the President of the Academy; the *papal address*, delivered by the pope to the Academy's scientists in response (at least seemingly) to the summary report just received from the President; and the genre of the *papal encyclical*. We found a high degree of intertextuality of content and language – patterns of uptake and recontextualization – binding these four genres together. For example, the focus of an ACTA, the summary of the discussions that have occurred among the Academy's scientists during a week-long biennial plenary meeting may reflect earlier prompts by a pope, in a previous papal address, to investigate a particular science-related topic in a subsequent plenary meeting, while papal encyclicals draw on prior papal addresses, among other sources.

In what follows, we present the body of findings from our study in two parts. We begin by situating the discursive representation of science/technology as a key element in Pope Francis's multi-faceted "integral ecology," a concept featured centrally in *Laudato Si'*, and then tracing a genealogy of this representation of science/technology into the past to show how the representation was derived from a decades-long dialogue between eight successive popes and generations of Academy scientists – on our part, an analytical task of discursive and epistemic "reverse-engineering," so to speak. In presenting results related to this genealogy, we describe the role of a genre set that has served as a discursive medium for the long-running dialogue, an ongoing exchange of knowledge and beliefs between Academy scientists and eight popes, reflecting their shared concern for the direction of contemporary science/technology.

Second, we describe another genre set, one used to communicate *Laudato Si'* to as broad and diverse an audience as possible. We then examine the numerous digital responses evoked by *Laudato Si'* following its release in June 2015, focusing on, given our emphasis in this chapter on representations of science/technology, those responses whose authors have taken up and recontextualized such representations for their own particular purposes. Finally, we examine the affordances (and one liability) inherent in the emergent digital version of the papal encyclical genre, as it has been evolving from its traditional print antecedent genre. Here we characterize *Laudato Si'* as an instance of a hybrid genre, one combining certain features of digital discourse with valued residual features of an antecedent genre and its traditional print text.

In considering the genealogy of the representation of science/technology inherent in *Laudato Si'*, describing the digital "launch-day" genre set used in the release of the *Encyclical*, and then examining digital textual responses to the *Encyclical*, we are tracing the trajectory of an intertextual chain of discourse and knowledge construction – more specifically, a discursive and epistemic chain of successive uptakes and recontextualizations. This trajectory of discourse originates in the post-1916 dialogue between Academy scientists and a succession of popes, a dialogue conducted largely through the ACTAs and papal addresses; and continues through to the digital responses elicited by *Laudato Si'* following its online appearance in June 2015.

5.1 The representation of science/technology in *Laudato Si'*: A genealogy

In our analysis of some sixty papal addresses, delivered by eight different popes at the conclusion of biennial plenary meetings dating from 1917 to 2015, we identified four recurrent themes that, considered as a whole, can be seen to form a collective argument (Smart 2011) – that is, as mentioned earlier, an argument developed and communicated by a group of social actors functioning together rhetorically – that advanced the Vatican's claim that humanity's use of science/technology must be instilled with morality, dedicating itself to the well-being of the Earth and all its inhabitants; and that otherwise, our embrace of science/technology will not end well, neither for humanity nor for the natural environment.

The four recurrent themes that emerged from our analysis are presented below. Each of the themes is illustrated with an excerpt from Pope Francis's *Laudato Si'*, followed by an excerpt from each of two former popes, with these appearing in reverse chronological order. With each theme, we can see the enactment of uptake and recontextualization, with the discourse of Pope Francis reflecting aspects of the discourse of the earlier pontiffs.

1. For the well being of humanity and the Earth, the Church, with its discourse of faith and morality, must maintain an ongoing dialogue with science/technology and their discourse of rationality. This synthesis of discourses will make possible the most productive and meaningful form of human thought and action. (italics and bolding added to the excerpts below)
   (Pope Francis, *Laudato Si'*, June 2015) *[S]cience and religion, with their distinctive approaches to understanding reality, can enter into an intense* ***dialogue*** *fruitful for both. ... The Catholic Church is open* ***to dialogue*** *with scientific thought; this has enabled her to produce various syntheses between faith and reason.*

(Pope Benedict XVI: Address to the Academy, November 2012) I am convinced of the urgent need for continued **dialogue** and cooperation between the worlds of science and of faith in building a culture of respect for man, for human dignity and freedom. … Without faith and science informing each other, the great questions of humanity … are abandoned to the irrational, to myth, or to indifference, with great damage to humanity itself, to world peace and to our ultimate destiny.
(Pope Paul VI, Address to the Academy, October 1964) The Church does not fear the progress of science. She undertakes willingly a **dialogue** with the created world and applauds the wonderful discoveries that scientists are making in that world. Every true scientist is for her a friend.

2. The Church recognizes, and indeed celebrates, the many benefits that science/technology have brought to humankind.
   (Pope Francis: *Laudato Si'*, June 2015) *We must be grateful for the praiseworthy efforts being made by scientists and engineers dedicated to finding solutions to man-made problems. … We are the beneficiaries of two centuries of enormous waves of change: steam engines, railways, the telegraph, electricity, automobiles, aeroplanes, chemical industries, modern medicine, information technology and, more recently, the digital revolution, robotics, biotechnologies and nanotechnologies.*
   (Pope Benedict XVI: Address to the Academy, November 2006) If we think … of how modern science, by predicting natural phenomena, has contributed to the protection of the environment, the progress of developing nations, the fight against epidemics, and an increase in life expectancy, it becomes clear that there is no conflict between God's providence and human enterprise.
   (Pope John Paul II: Address to the Academy, November 1979) In its other aspect, science turns to practical applications, which find their full development in the various technologies. In the phase of its concrete achievements, science is necessary to mankind to satisfy the rightful requirements of life, and to overcome the different ills that threaten it.
3. At the same time, however, science/technology have demonstrated their capacity to do great harm both to humankind and to the Earth.
   (Pope Francis: *Laudato Si'*, June 2015) *To this point in history, science/technology have been as harmful to humankind and to the natural environment as they have been beneficial. … [N]uclear energy, biotechnology, information technology, knowledge of our DNA, and many other abilities which we have acquired, have given us tremendous power. … [Y]et nothing ensures that it will be used wisely, particularly when we consider how it is currently being used, [including] the increasingly deadly arsenal of weapons available for modern warfare.*

(Pope John Paul II: Address to the Academy, November 1979) [It] is all but inevitable that a certain form of scientific research will be used for purposes of aggression. … [O]ften, the ever more sophisticated and deadly technology that has stemmed from science has been used against man, to the point of creating fearsome stockpiles of both conventional and nuclear arms, and of biological and chemical weapons, capable of destroying a large part of humanity.
(Pope Paul VI: Address to the Academy, April 1966) [W]e ought to face the … immense problem of the morality of the applications of science. Whether it is a question of genetics, of biology, of the employment of atomic energy, of many other fields which affect what is essential in man, the upright scientist cannot but question himself as to the bearing of his discoveries on [humankind]. Is everything permissible? Can applied science dispense with a norm of morality, can it proceed unchecked "beyond good and evil"? Who can fail to see the aberrations in which some could indulge in the name of science?

4. Given the sequence of statements above, science/technology, with their Janus face of beneficial and harmful effects, must take counsel from religious faith and morality and endeavor to function justly for the greater good of all people and the planet.
(Pope Francis: *Laudato Si'*) *[We face] an urgent need for a radical change in the conduct of humanity [given that] the most extraordinary scientific advances, the most amazing technical abilities, the most astonishing economic growth, unless they are accompanied by authentic social and moral progress, will definitively turn against man. … Here I would recall the balanced position of Saint John Paul II, who stressed the benefits of scientific and technological progress … [W]hile also noting that "we cannot interfere in one area of the* ***ecosystem*** *without paying due attention to the consequences of such interference in other areas." … We need constantly to rethink the goals, effects, overall context and ethical limits of this human activity.*
(Pope Benedict XVI: Address to the Academy, November 2010) [S]cientific achievement in this new century should always be informed by the imperatives of fraternity and peace, helping to solve the great problems of humanity, and directing everyone's efforts towards the true good of man and the integral development of the peoples of the world. The positive outcome of twenty-first century science will surely depend in large measure on the scientist's ability to search for truth and apply discoveries in a way that goes hand in hand with the search for what is just and good.
(Pope John Paul II: Address to the Academy, November 1979) There is no doubt that applied science has rendered and will continue to render immense services to man, provided it is inspired by love, regulated by wisdom, and accompanied by the courage that defends it against the undue interference of all

> tyrannical powers. Applied science must be united with conscience, so that, in the trinomial, science-technology-conscience, it is the cause of man's real good that is served.

As mentioned above, these four themes, taken jointly, can be seen to comprise a logical sequence of steps in a collective argument advanced, collectively, by the eight popes serving from 1917 to 2015 in support of the claim that science/technology, in order to be of most benefit to all of humankind and to the Earth, must temper the power of its rationality with a sense of morality and social justice.

Our analysis also identified another discursive pattern in the papal addresses delivered to the Academy's scientists by successive popes over the decades from 1917 to 2015: a growth in the ideational complexity of the representations of science/technology, a complexity that eventually found expression in the discourse of *Laudato Si'*. This increasing complexity includes three aspects, each illustrated below with an excerpt from *Laudato Si'* followed by an excerpt from either one or two earlier popes. Again, we see enactments of uptake and recontextualization within the discourse.

1. A trend away from a realist, low-agency representation of science/technology and the knowledge of reality they yield, towards a social constructionist representation of reality and scientific knowledge.
   (Pope Francis: *Laudato Si'*, June 2015) *Given the complexity of the ecological crisis and its multiple causes, we need to realize that the solutions will not emerge from just one way of interpreting and transforming reality. If we are truly concerned to develop an ecology capable of remedying the damage we have done, no branch of the sciences and no form of wisdom can be left out, and that includes religion and the language particular to it.*
   (Pope John Paul II: Address to the Academy, October 1990) The Church highly values [the Academy's] specialised research, which includes epistemological reflection on the meaning of science. Your studies bear witness to the efforts of human reason to explore reality more fully and to discover the truth in all its dimensions.
   (Pope Paul VI: Address to the Academy, April 1966) [Science] is a servant in relation to man, who is king of creation. If it were to refuse to serve, if it no longer aimed at the good and the progress of humanity, it would become sterile, useless and, let us say so, harmful. The consequences of this mission of service are incalculable, and here we ought to face … the immense problem of the morality of the applications of science. Whether it is a question of genetics, of biology, of the employment of atomic energy, of many other fields which affect what is essential in man, the upright scientist cannot but question himself as to the bearing of his discoveries on this psycho-physiological complex which is, in a word, a human person.

2. As part of this increasing social constructionist orientation, a recognition of the influence of science/technology on human consciousness, in both the individual and society.
   (Pope Francis: *Laudato Si'*, June 2015) *It can be said that many problems of today's world stem from the tendency, at times unconscious, to make the method and aims of science and technology an epistemological paradigm which shapes the lives of individuals and the workings of society. … [This] technological paradigm has become so dominant that it would be difficult to do without its resources and even more difficult to utilize them without being dominated by their internal logic. … Technology tends to absorb everything into its ironclad logic. … The effects of imposing this model on reality as a whole, human and social, are seen in the deterioration of the environment. … Life gradually becomes a surrender to situations conditioned by technology, itself viewed as the principal key to the meaning of existence.*
   (Pope Benedict XVI: Address to the Academy, November 2006) This increasing "advance" of science, and especially its capacity to master nature through technology, has at times been linked to a corresponding "retreat" of philosophy, of religion, and even of the Christian faith. Indeed, some have seen in the progress of modern science and technology one of the main causes of secularization and materialism.
3. A growing awareness of the power relations inherent in the control of science/technology by a social elite possessed of wealth and influence.
   (Pope Francis: *Laudato Si'*, June 2015) *[N]uclear energy, biotechnology, information technology, knowledge of our DNA, and many other abilities which we have acquired, have given us tremendous power. More precisely, they have given those with the knowledge, and especially the economic resources to use them, an impressive dominance over the whole of humanity and the entire world. Never has humanity had such power over itself, yet nothing ensures that it will be used wisely, particularly when we consider how it is currently being used … [including] the increasingly deadly arsenal of weapons available for modern warfare.*
   (Pope John Paul II: Address to the Academy, October 1992) Unarmed prophets have been the object of derision in every age, especially on the part of shrewd politicians, the supporters of power. But today must not our civilisation recognise that humanity has need of them? Should not they alone be heard by the whole of the world's scientific community, so that the laboratories and factories of death may give place to laboratories of life? … When, in a particular historical situation, it is all but inevitable that a certain form of scientific research will be used for purposes of aggression, he must make a choice that will enable him to work for the good of people, for the building up of peace.

> **(Pope John Paul II: Address to the Academy, November 1979)** There is no doubt that applied science has rendered and will continue to render immense services to man, provided it is inspired by love, regulated by wisdom, and accompanied by the courage that defends it against the undue interference of all tyrannical powers. Applied science must be united with conscience, so that, in the trinomial, science-technology-conscience, it is the cause of man's real good that is served.

5.2 *Laudato Si'*: The encyclical genre as part of two genre sets

Here we begin by examining what we will refer to as the "launch-day" genre set, which was used by the Vatican to extend the reach of *Laudato Si'* for its official release on June 18, 2015. This is a second genre set, in addition to the science/technology genre set discussed earlier, in which the *papal encyclica*l formed a pivotal part. Following this, we turn to tweets produced through the Vatican's Twitter account, viewing these tweets as part of the "launch-day" genre set and pointing to one of the genre's affordances. Next we consider the host of Internet-posted textual responses evoked by *Laudato Si'* following its release by the Vatican, focusing on those responses that took up and recontextualized some aspect of the representation of science/technology contained in *Laudato Si'*. Finally, we look at *Laudato Si'* as an instance of an evolving hybrid genre displaying both the features of a digital genre and certain valued aspects of the antecedent genre of the *encyclical*, with its traditional print text. Finally, we discuss an essential affordance of the digital *Laudato Si'* as well as a liability.

5.2.1 *The launch-day genre set*

Upon the Vatican's release of Pope Francis's *Laudato Si'*, the *Encyclical* was disseminated digitally worldwide via the launch-day genre set. This genre set created a discursive synergy among its constituent elements: a *Laudato Si'* homepage (https://laudatosi.com/) with live links to an online version of the "Official Vatican Text" available in a dozen languages, to a downloadable pdf of the English-language text, and to an online Amazon.com order form for purchasing the book version of the *Encyclical*; a six-minute video from the Vatican conveying key aspects of the text, in both English and Italian (https://www.youtube.com/watch?v=KXA5_juFgDg); a second video in which an American bishop comments on the Pope's integral ecology as a central part of *Laudato Si'* (https://www.youtube.com/watch?v=zWEK8JXQo0M); a third, animated, video, aimed at young viewers (https://vimeo.com/135230504); and a series of 63 tweets sent out through Pope Francis' Twitter account (@Pontifex). Functioning as a whole, this genre set elicited an immediate and widespread reaction to *Laudato Si'* in the public space of the Internet.

The launch-day genre set, in which the *encyclical* is again a focal element, continues the intertextual chain of uptake and recontextualization begun with the "science/technology" genre set described earlier. Below we will illustrate this uptake and recontextualization of *Laudato Si*"s representation of science/technology with six tweets sent out from the Pope's Twitter account on June 18, 2015:

Pope Francis @Pontifex · 18 Jun 2015
Scientific and technological progress cannot be equated with the progress of humanity and history.
160 2.3K 4.0K

Pope Francis @Pontifex · 18 Jun 2015
A true "ecological debt" exists, particularly between the global north and south.
85 2.0K 3.2K

Pope Francis @Pontifex · 18 Jun 2015
Climate change represents one of the principal challenges facing humanity in our day. #LaudatoSi
392 7.2K 7.8K

Pope Francis @Pontifex · 18 Jun 2015
Reducing greenhouse gases requires honesty, courage and responsibility. #LaudatoSi
271 3.1K 5.1K

Pope Francis @Pontifex · 18 Jun 2015
We need only to take a frank look at the facts to see that our common home is falling into serious disrepair. #LaudatoSi
142 2.5K 3.7K

Pope Francis @Pontifex · 18 Jun 2015
The human environment and the natural environment deteriorate together.
172 5.8K 7.3K

These tweets involved repeated acts of uptake and recontextualization where aspects of *Laudato Si*"s representation of science/technology were transformed and repurposed by the Vatican in new texts, either explicitly or implicitly, and either word-for-word or as paraphrase. Close attention to the sixty-three tweets posted by the Vatican, particularly the twenty tweets focusing on science/technology, reveals that these tweets jointly advance an aspect of the same collective argument discussed above: that humanity's relationship with science/technology must be imbued with morality and dedicated to the well-being of the Earth and all its inhabitants, or our embrace of science/technology will end in disaster, both for humankind and for the natural environment.

Twitter has several affordances that together allow tweets to generate, almost immediately after being posted, large numbers of related comments and new posts, creating a "digital conversation" – which is exactly what occurred when the Pope's Twitter account began to post tweets on June 18, 2015: the tweets evoked a multitude of responses, including a number focusing on *Laudato Si'*'s representation of science/technology. At different times during the day of *Laudato Si'*'s release the Vatican had as many as 2,300 individuals involved in this digital conversation.

5.2.2 *Internet-posted textual responses to* Laudato Si'

The many thousands of Internet-posted responses to the *Encyclical* appeared in various genres – including blog posts, newsletters, e-bulletins, tweets, news reports, educational materials, press releases, and academic articles – originating from a wide span of interest groups, institutions, and individuals, such as Church congregations, business associations, news media, NGOs, think-tanks, partisan political alliances, and religious organizations. Unsurprisingly, given the breadth of content included in *Laudato Si'*, these Internet-posted responses included a range of discourses (Hulme 2009): scientific, cultural, economic, political, and social-justice discourses. In a number of the responses the main theme was science/technology, with the authors taking up and recontextualizing *Laudato Si'*'s representation of science/technology for their own ideological ends. In some cases this was done to support the Pope's plea for an integral ecology, such as in the example below:

> [T]he pope proposes that we need to acknowledge that science is not enough. We need to acknowledge other sources of wisdom, religion included, to help us to appreciate the ends that make life meaningful. … Strikingly, therefore, Pope Francis recommends the reading of religious classics alongside the doings of science … [A]ccording to the pope, science alone cannot provide a comprehensive [answer] to our common social and ecological problems because it will be unable to move people to undertake the self-denial that will be required of us to live in harmony with creation. (Austriaco 2015)

In other contrasting cases, the uptake and recontextualization of *Laudato Si'*'s representation of science/technology resulted in a text that attacked the Pope and his vision of an integral ecology, as we see here:

> Having read through Pope Francis' new encyclical, I am dismayed at how many groundless assertions it makes. From a strictly scientific point of view, *Laudato Si'* is an embarrassment. I'm not [just] talking about the much ballyhooed support the encyclical gives to climate change. The problem [also] lies in other questionable claims it makes about the "environmental crisis," and "the symptoms of sickness evident in the soil, in the water, in the air, and in all forms of life." Pope Francis is not trained in the natural sciences and so necessarily relied on environmental "experts" selected by others to advise him on these matters. ... The people pushing this agenda aren't interested in smart solutions that benefit humans while protecting the planet. They're alarmists who won't be happy until we're back in caves. (Mosher 2015)

5.2.3 *Laudato Si'*: An instance of a hybrid genre

We view *Laudato Si'* as an instance of a hybrid genre. While it displays certain primary features of digital discourse, it also retains valued characteristics of the antecedent *encyclical* genre with its traditional print text. The most obvious of these characteristics can be seen in the format of the first page of the encyclical (see page 214).

As we can see, the hybrid version of the *Encyclical* retains the visual and structural design of the older print version. We would speculate that the reason for this is the Vatican's wish to retain the *ethos* of the traditional print version of the genre, while at the same time benefiting, rhetorically, from the affordance of an immediate and worldwide digital dissemination of *Laudato Si'*.

Another key difference between the digital version of *Laudato Si'* and its traditional print antecedent concerns their respective audiences. While the latter was limited, relatively narrowly, to a readership comprising members of the Catholic Church, *Laudato Si'* had greater ambitions, given its rhetorical purpose, as indicated by Pope Francis:

> Now, faced as we are with global environmental deterioration, *I wish to address every person living on this planet.* In my *Apostolic Exhortation Evangelii Gaudium* [another form of papal communication, lower in authority], I wrote to all the members of the Church with the aim of encouraging ongoing missionary renewal. *In this Encyclical, I would like to enter into dialogue with all people about our common home.* (Pope Francis 2015: 4) (italics added)

Finally, our analysis also points to an audience-related rhetorical liability of *Laudato Si'*. While the *Encyclical* benefited from the affordance of an immediate worldwide digital distribution of its message, this also gave critics of Pope Francis and his integral ecology a public forum for judging and, in many cases, condemning the message. This liability was evident with Twitter, for example, where we saw a number of antagonistic responses to Pope Francis' message.

ENCYCLICAL LETTER

LAUDATO SI'

OF THE HOLY FATHER

FRANCIS

ON CARE FOR OUR COMMON HOME

1. "LAUDATO SI', mi' Signore" – "Praise be to you, my Lord". In the words of this beautiful canticle, Saint Francis of Assisi reminds us that our common home is like a sister with whom we share our life and a beautiful mother who opens her arms to embrace us. "Praise be to you, my Lord, through our Sister, Mother Earth, who sustains and governs us, and who produces various fruit with coloured flowers and herbs".[1]

2. This sister now cries out to us because of the harm we have inflicted on her by our irresponsible use and abuse of the goods with which God has endowed her. We have come to see ourselves as her lords and masters, entitled to plunder her at will. The violence present in our hearts, wounded by sin, is also reflected in the symptoms of sickness evident in the soil, in the water, in the air and in all forms of life. This is why the earth herself, burdened and laid waste, is among the most abandoned and maltreated of our poor; she "groans in travail" (*Rom* 8:22). We have forgotten that we ourselves are dust of the earth (cf. *Gen* 2:7); our very bodies are made up

(Pope Francis 2015: i, 1)

6. Conclusion

In concluding we make a claim for the value of our research and then point to a related path of future inquiry. The primary significance of our findings is three-fold. First, we have shown how the science/technology genre set served as a discursive vehicle supporting an intertextual chain of uptake and recontextualization that has been used, for almost a century, to develop an evolving body of institutional knowledge and collective argumentation regarding science/technology and its place in the world, knowledge and argumentation that eventually emerged in

Laudato Si''s representation of science/technology. Second, we have demonstrated how another genre set, the launch-day genre set, was employed to digitally disseminate *Laudato Si'* around the world within a short time-frame. With each of these digital genre sets, we revealed the interdependence among their constituent genres, while at the same time showing how the two genre sets are themselves interconnected, with *Laudato Si'* linking the two. These genre sets can be viewed as functioning together as a medium for a continuous intertextual chain of uptake and recontextualization, regarding science/technology, that began early in the twentieth century as part of the dialogue between successive popes and generations of scientists in the Academy and concluded with, for our purposes here, the multitude of public responses to *Laudato Si'* following its release by the Vatican. We would hope that researchers in genre studies and perhaps in related fields of scholarship may find this conceptual model useful in their work, in whole or in part. Thirdly, our findings have led us to see *Laudato Si'* as a hybrid genre intentionally combining certain digital features and affordances with rhetorically valued aspects of the antecedent traditional print version of the genre. We hope this will be of interest to other researchers.

Looking ahead, we would like to investigate the discourse genres of another Vatican institution, the Pontifical Academy of Social Sciences, founded in 1994 by Pope John Paul II. Its mission runs parallel to that of the Pontifical Academy of Sciences, with a similar charge to meet biennially in Vatican City to confer on recent advances in the social sciences and to join with the current pope in continuing an ongoing dialogue about relevant topical developments in these fields of scholarship – particularly in economics, political science, and sociology. The focus in this dialogue is on the Church's tradition of 'social teaching', an institution with a mandate to promote the sanctity of human life and the dignity of all people in an effort to advocate for and eventually realize a moral vision for society. We believe that it would be most interesting to look at the genealogy of the genre set with which the pope and the members of the Pontifical Academy of Social Sciences have engaged as they have developed knowledge related to other aspects of *Laudato Si'* and the integral ecology of Pope Francis.

References

Aristotle. 1960. *The Rhetoric of Aristotle*. (L. Cooper, trans.). New York: Appleton-Century-Crofts.

Austriaco, Nicanor P. G. 2015. "The Relationship between Science and Religion according to '*Laudato Si'*'." *BioLogos*. Last accessed June 15, 2019. https://biologos.org/blogs/guest/the-relationship-between-science-and-religion-according-to-laudato-si

Bauer, Martin W., and George Gaskell. 1999. "Towards a Paradigm for Research on Social Representations." *Journal of the Theory of Social Behaviour* 29 (2): 163–186. https://doi.org/10.1111/1468-5914.00096

Bazerman, Charles. 1994. "Systems of Genres and the Enactment of Social Intentions." In *Genre and the New Rhetoric*, ed. by Aviva Freedman, and Peter Medway, 79–101. Bristol, PA: Taylor and Francis.

Bazerman, Charles. 1988. *Shaping Written Knowledge: The Genre and Activity of the Experimental Article in Science*. Madison, WI: University of Wisconsin Press.

Bhatia, Vijay K. 2016. "Genre as Interdiscursive Performance in Public Space." In *Genre and the Performance of Publics*, ed. by Mary Jo Reiff, and Anis Bawarshi, 25–42. Boulder, CO: Utah State University Press. https://doi.org/10.7330/9781607324430.c001

Billig, Michael. 1988. *Arguing and Thinking: A Rhetorical Approach to Social Psychology*. Cambridge: Cambridge University Press.

Burke, Kenneth. 1969. *A Rhetoric of Motives*. Berkeley: University of California Press.

Crowston, Kevin, and Marie Williams. 2000. "Reproduced and Emergent Genres of Communication on the World Wide Web." *Information Society* 16 (3): 201–215. https://doi.org/10.1080/01972240050133652

Devitt, Amy. 1991. "Intertextuality in Tax Accounting: Generic, Referential, and Functional." In *Textual Dynamics for the Professions: Historical and Contemporary Studies of Writing in Professional Communities*, ed. by Charles Bazerman, and James Paradis, 335–57. Madison, WN: University of Wisconsin Press.

Freadman, Anne. 2002. "Uptake." In *The Rhetoric and Ideology of Genre*, ed. by Richard Coe, Lorelei Lingard, and Tatiana Teslenko, 39–5. Cresskill, NJ: Hampton Press.

Graham, Scott, and Brandon Whelan. 2008. "Mode, Medium, and Genre: A Case Study of Decisions in New-media Design." *Journal of Business and Technical Communication* 22 (1): 65–91. https://doi.org/10.1177/1050651907307709

Grim, John. 2015. "President's Corner." *Teilhard Perspective* 48 (1): 1–5. Last accessed June 15, 2019. http://teilharddechardin.org/mm_uploads/TP_Spring_Summer_2015.pdf

Hajer, Marteen. 2006. "Doing Discourse Analysis: Coalitions, Practices, Meaning." In *Words Matter in Policy and Planning*, ed. by Margo van den Brink, and Tamara Metze, 65–74. Utretch, The Netherlands: Kovvinkliijk Nederlands Aardrijkskundig Genootschap.

Hall, Stuart. 1997. *Representation: Culture Representation and Signifying Practices*. London, UK: Sage.

Herring, Susan. 2013. "Discourse in Web 2.0: Familiar, Reconfigured and Emergent." In *Discourse 2.0: Language and New Media*, ed. by Deborah Tannen, and Anna Marie Trester, 1–25. Washington, DC: Georgetown University Press.

Hesse-Biber, Sharlene, and Patricia Leavy. 2008. *Handbook of Emergent Methods*. New York, NY: Guilford Press.

Hulme, Mike. 2009. *Why We Disagree about Climate Change*. Cambridge: Cambridge University Press. https://doi.org/10.1017/CBO9780511841200

Jamieson, Kathleen M. 1975. "Antecedent Genre as Rhetorical Constraint." *Quarterly Journal of Speech* 61: 406–415. https://doi.org/10.1080/00335637509383303

Linell, Per. 1998. "Discourse across Boundaries: On Recontextualizations and the Blending of Voices in Professional Discourse." *Texts* 18 (2): 143–157.

Luzón, María-José. 2013. "Public Communication of Science in Blogs: Recontextualizing Scientific Discourse for a Diversified Audience." *Written Communication* 30 (4): 428–457. https://doi.org/10.1177/0741088313493610

Mehlenbacher, Ashley R. 2019. "Registered Reports: Genre Evolution and the Research Article." *Written Communication* 36 (1): 38–67.
Miller, Carolyn R. 1984. "Genre as Social Action." *Quarterly Journal of Speech* 70: 151–167. https://doi.org/10.1080/00335638409383686
Miller, Carolyn R. 2016. "Genre Innovation: Evolution, Emergence, or Something Else?" *The Journal of Media Innovations* 3 (2): 4–19. https://doi.org/10.5617/jmi.v3i2.2432
Moscovici, Serge. 1963. "Attitudes and Opinions." *Annual Review of Psychology* 14: 231–260. https://doi.org/10.1146/annurev.ps.14.020163.001311
Mosher, Steven. 2015. "Do the Pope and I Live on the Same Planet?" *New York Post*. July 05. Last accessed June 15, 2019. https://nypost.com/2015/07/05/do-the-pope-and-i-live-on-the-same-planet
Potter, John. 1996. *Representing Reality: Discourse, Rhetoric and Social Construction*. London, UK: Sage. https://doi.org/10.4135/9781446222119
Smart, Graham. 2011. "Argumentation across Web-based Organizational Discourses: The Case of Climate Change." In *Handbook of Communication in Organisations and Professions*, ed. by Srikant Sarangi, and Christopher Candlin, 363–386. Berlin, DE: Mouton de Gruyter. https://doi.org/10.1515/9783110214222.363
Smart, Graham, and Matthew Falconer. forthcoming. "The Uptake and Recontextualization of Climate-change Science within 'Denialist' Cultural Communities." In *Genres of the Climate Debate*, ed. by Christel Sunesen, H. Bildsøe, and Sune Auken. Berlin, Germany: de Gruyter.
The Catholic Encyclopaedia, 2017. "Encyclicals." N.p.n. Nashville: HarperCollins.

Appendix: Corpus of analysis

Pope Benedict XVI. 2006, November 06. Address to the Pontifical Academy of Science.
Pope Benedict XVI. 2010, October 28. Address to the Pontifical Academy of Science.
Pope Benedict XVI. 2012, November 08. Address to the Pontifical Academy of Science.
Pope Francis. 2015. *Laudato Si', on Care for our Common Home*. Huntington, IN: Our Sunday Visitor.
Pope John Paul II. 1979, November 10. Address to the Pontifical Academy of Science.
Pope John Paul II. 1990, October 29. Address to the Pontifical Academy of Science.
Pope John Paul II. 1992, October 31. Address to the Pontifical Academy of Science.
Pope Paul VI. 1964, October 03. Address to the Pontifical Academy of Science.
Pope Paul VI. 1966, April 23. Address to the Pontifical Academy of Science.

CHAPTER 11

Public- and expert-facing communication

A case study of polycontextuality and context collapse in Internet-mediated citizen science

Gwendolynne Reid and Chris M. Anson
Oxford College of Emory University / North Carolina State University

This chapter describes a qualitative case study of digitally-mediated production and communication of research in the biological sciences. The study focuses on the citizen science "Heartbeats Project," conceived by a U.S.-based evolutionary biology lab to explore the data behind the well-known rule that, on average, mammals' hearts beat one billion times per lifetime. Our analysis describes three ways that polycontextuality and context collapse figured in the team's production of digital, spoken, and print-based genres arising from their work. These dynamics complicate traditional understandings of the relationships between scientific and public genres, as well as existing conceptions of composition, genre, authors, and audiences in the production and circulation of scientific findings and the (re)production of science.

Keywords: scientific communication, scientific writing, digital writing, digital genres, citizen science, public understanding of science, polycontextuality, context collapse

1. Introduction and theoretical framework

Until the advent of the Internet, scientific discoveries reached the public through conventional channels of print publication, lectures, and eventually television, often mediated by writers and editors who could translate sophisticated material into publicly consumable form. Today, the Internet affords opportunities for scientists to "disintermediate" their work, turning it "inside out" for more direct communication with public audiences (Trench 2008), as well as among members in various disciplines who do not always share the same specialized discourse.

Among scientific communities, there is broad consensus about the need for this disintermediation. Public engagement in science is "imperative" (Marincola

https://doi.org/10.1075/pbns.308.11rei
© 2019 John Benjamins Publishing Company

2006) and the "key for the future of science research" (Pham 2016), and should be a "normative commitment" (Stilgoe, Lock, and Wilsdon 2014). An American Association for the Advancement of Science survey reported by the PEW Research Center found that 87% of 3,748 scientists agreed with the statement "Scientists should take an active role in public policy debates about issues related to science and technology" (Rainy, Funk, and Anderson 2015). Yet in spite of these sentiments, the general public is still strongly dissociated from the inner workings of science. In an attempt to involve citizens in scientific inquiry and not just give them facts and information, new organizations such as the Citizen Science Alliance (http://citizenscience.org) and the Citizen Science Association (https://www.citizensciencealliance.org) are drawing on "distributed communities" of citizens as collaborators in the pursuit of scientific inquiry, partly to garner crowd-sourced information but also to promote broader understandings of science and respect for scientific inquiry.

Although considerable scholarship has explored both the theoretical dimensions and the practicalities of the relationship between scientists and the general public, we still lack a full understanding of the complex, contextually-situated processes involved in the creation and dissemination of scientific inquiry to public audiences and especially in their reciprocal participation and the role of different genres. In particular, the affordances of digital technology have dramatically changed the ways that scientific communication is carried out both within and beyond the specialized circles involved in its production, and the extent to which communicative genres have become "polycontextual" (Engeström, Engeström, and Kärkkäinen 1995; Zachry 2000), simultaneously realizing multiple purposes, reaching diverse audiences, populating manifold networks, and drawing in – and on – publics previously positioned as consumers of science rather than participants in its work.

The development of routine polycontextuality has helped to redefine genres and the contexts of their production. Earlier scholarship on writing after the "social turn" proposed the concept of the "discourse community," a "group of individuals bound by a common interest who communicate through approved channels and whose discourse is regulated" (Porter 1986: 38–39). The discourse community was said to determine or subsume the knowledge and processes required to create coherent texts, such as considerations of audience, style, structure, and other features of genre normed by "those bringing certain language-using practices" (Bizzell 2009: 222). Writers joining such communities had to "develop knowledge of genres whose boundaries and features the discourse community defines and stabilizes" (Beaufort 2007: 20).

Further research soon contested and complicated these assumptions, in part because it became clear that the "community" is always in a state of development and is less closed than it appeared. The stable, normative concept of the discourse community and its genres could not account for "individual roles and layers of knowledge as well as for the fluidity of 'community boundaries ...'" (Donahue 2017). In addition, the participants themselves did not represent a group of individuals bound by the same understandings of genres and habituated to the same practices; rather, they were in "dynamic flux ... entering, working through, and perhaps exiting a community as sometimes novice, sometimes expert, and sometimes in-between, depending on the activity, the stance, and so on" (Donahue 2017: 112). Yet in the context of scientific writing, as illustrated in Myers' study of science journal editors and scientists communicating with the public, experts were still assumed to be "translating" or "adapting their research" to reach more general audiences, who were still positioned as passive consumers of knowledge (Myers 1990: 191).

The advent of social-historical activity theory (see Russell 1995) also extended the concept of the discourse community through a richer understanding of context that included goals and motivations, tools and resources, and especially the ways that genres reflect layers of complex social practices in states of evolution. Instead of textual representations of static conventions, genres are reflections of dynamic processes within complex activity systems that are only temporarily stabilized. The idea that genres operate within (and mediate) specific contexts (meaning both physical and rhetorical spaces) soon gave way to an understanding that writing crosses rhetorical boundaries, adapting itself to multiple purposes and audiences. The spaces of writing in what appear to be stable communities are actually "conditional" (Anson and Dannels 2004), characterized by a network of activity involving diverse groups of people playing different roles and embodying different goals and purposes that sometimes are manifested in the text and sometimes only in what circumstances produce it (see Bazerman 1997).

Such conditions have always been present in the production and dissemination of print text, but the affordances of digital communication created an explosion of possibilities and practices, massively diversifying audiences and genres and allowing scholars to extend the reach of their work far beyond the boundaries of their intellectual circles. In their work on genres in the Internet, for example, Giltrow and Stein (2009: 9) observe that "the Internet enables a new communication *setting* which reconfigures conditions to which pragmatic features of language respond" (original emphasis). In their view, genres on the Internet have a "greater fluidity and pragmatic openness" (Giltrow and Stein 2009: 9–10), potentially "creat[ing] new functions, or new ways to execute established function." In particular, Internet-mediated communication has a capacity for "connect[ing] across contexts," affording opportunities for "new paths of interaction interlacing multiple contexts, traditional and new" (Giltrow and Stein 2009: 22). They conjecture that changes to the "epiphenomena of the technical and communicative setting" – changes in users' perceptions and experiences of, or the "psycho-emotional dimensions" of, Internet-mediated communication – are as important to understanding the impact of the Internet on genre and communication as changes inherent to the media themselves (Giltrow and Stein 2009: 13). For many Internet users, one change may be "a lesser sense of 'domination,'" or greater sense of communicative freedom, "either in terms of topic or in terms of social relationships, or in terms of the kinds of parameters which have traditionally given rise to discourse communities" (Giltrow and Stein 2009: 13). In their discussion, however, Giltrow and Stein classify science, and scholarly communication more broadly, as a case apart, positing that the scientific research article may not experience the same sorts of radical generic changes as genres in other contexts "if the uptake of the speech act" and the "activities of reception and response" remain similar to those "of the print form of the scientific article" (Giltrow and Stein 2009: 13). We know, however, that like other Internet users, scientists have experienced new media as presenting possibilities for science, including creating reciprocal channels of communication with previously unresponsive audiences, such as through citizen science (Davies and Hara 2017). The expansion of opportunities for this reciprocity has raised new questions about the functions of discourse within specific projects involving citizen science, because earlier conceptions of the "popularization" of scientific findings to lay audiences have given way to a situation in which many genres are simultaneously enmeshed with each other.

The study reported here extends contemporary approaches to genre and digitally-mediated communities. In particular, we are interested in what Davis and Jurgenson (2014) call "context collapse," or the meshing of social contexts both consciously enabled ("context collusions") and tacitly allowed ("context collisions") because of online digital communication. Giltrow and Stein (2009: 11) posit that "the functions of Internet genres change often" and exhibit a "low degree of conventionalization of form-function relationships," making them "amenable more to pragmatic-functional than to linguistic description." In keeping with this approach and rhetorical studies of genre, this study offers a phenomenological rather than linguistic account of a scientific team's experiences with the pragmatic conditions offered by context collapse, which included routine public communication as part of citizen science. Based on a text-focused ethnography of a scientific team's digitally-mediated production of a research article for the scientific community but stemming from a citizen science project, the study provides thick description of the scientific authors' production of digital, spoken, and print-based genres and the multiple social scenes simultaneously interlaced as part of this production. These dynamics underscore the complexly enmeshed communication settings of digital scientific genres and complicate traditional understandings of the relationships between scientific and public genres, pointing toward routine polycontextuality in the production and circulation of scientific findings and the (re)production of science.

2. Research context

The study described in this chapter examined the scientific writing practices of a biology lab at a public research university in the Southeastern United States. Initially classified as an entomology lab in 2013, it transitioned to a microbiology lab by the end of the study period in 2017. As Hine (2006: 273) remarks about scientific laboratories more generally, it consisted of "more than a physical space," including a floor of a scientific building, space in two forests owned by private universities, webspace on a lab website and blog, and an extended social network that included lab alumni and collaborators at other institutions. This distributed social network was enabled by and required both physical interaction and extensive use of digital communication, including an assortment of genres (a lab listserv, blog, and social media accounts such as Twitter and Facebook). In addition to digitally-mediated communication, the lab's multidisciplinary and cross-institutional collaboration was also supported by strategies such as organizing workshops or "think tanks" as occasions to invite collaborators to travel to a given location (see Cummings and Kiesler 2005). The lab was headed by Soren,[1] an evolutionary biologist with a commitment to public science, a term he used for engaging the public with science both through traditional outreach and involvement in the scientific process itself. As part of this commitment, Soren had received a large, multiyear National Science Foundation (NSF) grant for developing middle and high school science lesson plans that enabled students to contribute their resulting data to professional scientific research projects. To facilitate the grant, the lab ran a range of citizen science projects to which anyone, including science classrooms, could contribute data. These projects, many of which ended up with thousands of citizen science contributors, were an opportunity to engage students and members of the public in the process of "doing science," as members of the lab put it, and then later with the results of the studies.

As part of her doctoral dissertation, in the fall of 2014 Gwendolynne began an ethnographic study of the lab's digital scientific writing focused on the text trajectory (Lillis and Maybin 2017) of a developing manuscript. With knowledge of Gwendolynne's project, Soren matched her with a lab citizen science project and corresponding manuscript still in its early stages, which together we are calling the

1. Participant and project names are pseudonyms.

"Heartbeats Project." The lead author on this manuscript was a postdoctoral fellow, Clay, whose area of expertise was the evolution and ecological consequences of behavior in social insects. The project had its beginnings in a popular science book Soren had written and that was being published that spring, continuing his regular authorship of books and blog posts for the public.

Soren had asked Clay and a research technician to research the "billion beats rule," which is based on the finding that mammals, on average, experience one billion heartbeats per lifetime, with some species' hearts beating quickly (leading to short lives) and others' beating slowly (leading to longer lives). They soon discovered that this rule was largely based on a 1997 article by cardiologist Herbert Levine, who had examined the resting heart rate and life expectancy of 16 species and extrapolated from this that mammals had an average of a billion heartbeats per life. However, humans seemed to be an exception, with a substantially higher number of beats per life. As they examined this limited data, Soren and his team felt that the study merited replication with more extensive and precise data.

The Heartbeats citizen science project was conceived as a way to collect data to test the billion beats rule, while engaging citizen scientists in finding, reading, and evaluating scientific literature. In this way, the project involved a kind of reciprocity: unlike many public science initiatives designed to "translate" complex scientific findings or discoveries to passive lay audiences, the Heartbeats Project solicited the engagement and contributions of the public, potentially enhancing their connection with and interest in the science. Through a set of online instructions, citizen scientists were invited to choose a species not yet represented in the data and locate resting heart rate and longevity data using several suggested scientific databases. They could then submit the data to the team for checking and integration into the study. Gwendolynne began ethnographic engagement with the project as the team rolled out the citizen science page and as Soren's book was launched.

3. Design and methods of the study

This study was designed as a text-oriented ethnography of a scientific text's trajectory (Lillis and Maybin 2017), focusing on the development, between 2014 and 2017, of a scientific manuscript intended for digital publication. Gwendolynne's ethnographic engagement over these three years sought to close the gap between participants' text and context in order to "build holistic understandings" of their scientific writing world and its genres (Lillis 2008: 362), an approach that can make the pragmatic-functional dimension of text for participants visible (Giltrow and Stein 2009: 11). Beginning in 2014, Gwendolynne combined periodic observations of team and lab meetings and author interviews focused on drafts and process-related artifacts with mediated engagement with lab digital communication (listservs, Twitter and Facebook feeds, blog posts, website updates, etc.) and digital tools and resources (databases, spreadsheets, software, journal websites). To deepen her understanding of what was "significant and at stake" for her participants at that moment (Lillis 2008: 367), Gwendolynne included "thick participation" in the team's scientific world as part of her ethnographic engagement. This included participation in the Heartbeats Project as a citizen scientist, as well as participation in public events related to the project, other lab citizen science projects, and citizen science events beyond the lab (#CitiSciChat Twitter events, museum-sponsored citizen science events, citizen science databases and blogs). "Indexicality," or "the specific ways in which bits of language (speech, writing) index, or point to aspects of the social context" (Lillis 2008: 376) guided Gwendolynne's use of participant talk and writing to explore what project members found to be significant elsewhere in their sociocultural context.

Beaulieu (2010) notes the challenges that distribution and mediation present for researchers who study scientists, offering *co-presence* as a strategy for reconceiving the field beyond a physical lab and emphasizing this is an "active form of 'field-making'" with the field "constituted *in* the interaction," including mediated interaction (463). Following researchers such as Gupta and Ferguson (1997) in anthropology and Hine (2015) in studies of e-science, Gwendolynne therefore made "location work" central to her ethnographic engagement with the team's scientific writing (Gupta and Ferguson 1997: 5), including thick participation in both physical and digital sites significant to the team and project and integrating these as part of her co-constructed field through field notes, analytic memos, and artifact collection (e.g. screenshots of digital and physical promotional materials for citizen science events). Disciplines have long been (re)produced through mediated interaction, a fact Gwendolynne accounted for by purposively engaging with print culture in the form of citations, articles, and books (Hine 2015: 133), as well as digital resources like the AnAge database used in the Heartbeats Project.

Data analysis was informed by grounded theory (Glaser and Strauss 1967), specifically the constant comparative approach, with data collection and analysis each informing the other (Charmaz 2003). Collected data included field notes and analytic memos based on Gwendolynne's engagement with the emergent field, as well as field-related artifacts and genres, such as interview transcripts, photographs, screenshots, PDFs, and Storified Twitter events, and process-related artifacts, such as drafts, spreadsheets, figures, and sources (see Table 1). These were collected and coded using the digital qualitative data analysis tool MAXQDA and a cyclical process of open coding and then focused coding (Charmaz 2003; Saldaña 2009), in order to develop categories that reflected participants' experiences of developing their manuscript within their sociocultural world. Coding was "sensitized" (Charmaz 2003) by concepts such as genre, medium, mode, and discipline, but was also conducted inductively to develop meanings significant to participants. In addition to qualitative coding, process-related artifacts such as drafts were analyzed for their rhetorical moves, with these text-related insights put into conversation with ethnographic insights through discourse-based interviews (Odell, Goswami, and Herrington 1983) and analytic memoing (Saldaña 2009). In addition to traditional scientific research genres, such as research articles, conference presentations, and grant proposals (see Swales 2004), the team's genre repertoire included digital genres such as research blog posts, citizen science tweets, citizen science online instructions, and citizen science project webpages.

Table 1. Data collection within and beyond Heartbeats Project team

Within team	Beyond team
16 Interviews (18 hours)	Citational Connections (204 articles)
10 Observations (20 hours)	SciStarter Observations (4 projects)
Drafts & Writing Samples (12 texts)	Database Observations (e.g. AnAge) (5 databases)
Project Data Spreadsheets (20 spreadsheets)	Digital Tool Observations (5 tools)
Project Website/Lab Blog Posts (14 posts)	Journal Submission Guidelines (5 texts)
Emails (262 threads)	Twitter #CitiSciChat Events (3 events)
CVs & Published Profiles (4 texts)	CitSci Listserv Observations (270 listserv posts)
Media Coverage of the Project/Lab (12 articles)	Citizen Science Participation (3 projects)
Lab Social Media Account Subscriptions (3 platforms)	Local Citizen Science Events (3 events)
Participation in the Heartbeats Project (1 submission)	NSF Documents and Databases (4 documents)
	Citizen Science Association Website and Blog (2 site visits)

The present study focuses on three dimensions that emerged as relevant to understanding the Heartbeats team's digital genre performances: (1) *public communication as scientific invention*; (2) *composing for recomposition*; and (3) *composing for/with the public to change science*. In the context of this collection, these three dimensions demonstrate the ways in which the processes of creating text are continuing to evolve through systemic changes in the development and role of genres, especially as these are mediated through new and emerging technologies that "reconfigure" features of language (Giltrow and Stein 2009), and the ways in which more complex, layered audiences and participants are creating new reciprocal methods of scientific inquiry, communication, and understanding. Ethical procedures established in collaboration with the university institutional review board office were followed, including the use of pseudonyms and strategic disguising of the Heartbeats Project to protect participant confidentiality (Bruckman 2002).

4. Analysis of the case

In the following sections, we describe and document the three aforementioned ways that polycontextuality and context collapse affected the processes that the Heartbeats Project team used in the production of genres arising from their work. Each observation pressures existing conceptions of composition, genre, and the role of audience in the production and circulation of scientific information. These, for example, complicate traditional understandings of the sequences between published scientific genres, genres related to the public communication of science, and process-related texts and genres for both. The compositional choices made by writers as part of these altered sequences also pressure notions of the roles, agency, and make-up of public and scientific audiences, ultimately extending our understanding of scientific discourse in the digital age.

4.1 Public communication as scientific invention

To understand the polycontextuality at work in the Heartbeats Project team's digital scientific writing, we begin with the role of public audiences in the invention for that writing. As described earlier, the study arose from Soren's work on a popular science book. While researching the scientific literature for a line of thinking in this book, the team discovered a gap that they felt would be beneficial and feasible for them to fill. According to Clay and Summer, the two members of the lab who helped Soren in his literature search, the manner of this discovery was not unusual for Soren's lab. In an early interview, Summer described Soren as writing quickly and following ideas with intensity, often noticing gaps and asking his team to locate data to fill them, with those gaps sometimes becoming new projects if they couldn't find any relevant data or studies. Rather than seeing the activity of producing one genre (a popular text) as separate from and secondary to the activity of producing scientific meaning through other genres (scientific research articles), Soren and his lab members framed this for themselves as part of the same activity of "doing science." The widely understood notion that writers, in an act of invention, write in order to think was not unusual for the Heartbeats team. Yet we tend to conceptualize this thinking as contextually focused, typically to produce a single text, following a tradition that stretches back to classical rhetoric (see Lauer 2004). In this case, however, Soren and his team used the thinking generated by their writing for a *public* context in order to advance thinking and writing for a *scientific* context. Rather than a unique, serendipitous occurrence for the lab, this relationship was planned for and part of how their research and writing genre ecosystem routinely functioned, with different genres mutually influencing each other.

This use of public science writing as part of the team's scientific invention is not distinctly digital – locating research gaps while writing popular science likely occurred in earlier media environments. A closer look, however, reveals how the team's process was inextricably tied to contemporary digital environments. Soren's initial request to Clay and Summer for research assistance was sent as an email that led the team to set up a shared online folder on Dropbox to coordinate spreadsheets and article PDFs as they pursued this fresh line of inquiry using Google Scholar and several online databases and data repositories (see Table 2 for a list of digital tools used). Later, they created the first draft of their scientific research article manuscript online in Google Docs, which mediated a highly collaborative process with geographically distributed members of the lab. While Clay created the subsequent draft of the manuscript in a more solitary manner, he did so using materials gathered from multiple digital genres and platforms (Google Docs, the citizen science webpage, abstracts in other Word files, etc.) into a single Word file. In Clay's opinion, the very genre of their manuscript – a meta-analysis – was enabled by the Internet, beginning with the initial search in Google Scholar for the keyword "heartrate" and then the team's research journey across article PDFs, online databases, and data repositories to locate pre-existing data for repurposing in their analysis. The evocatively-named tool, Data Thief, was in fact designed to enable this sequence of research genres, helping scientists "extract" data from research articles and data repositories to be repurposed in meta-analyses, an increasingly common research genre. The polycontextuality at work in Soren's inquiry process, then, while not unique to the contemporary era, may not have been possible in the same routinized way as evident in his lab's work.

The use of public communication for scientific invention extended beyond the relationship between Soren's popular science book and the manuscript's research gap. The team's citizen science project prompted extensive public communication in a range of genres. Most of these were designed to provide reciprocal channels of communication with citizen scientists. This communication included a project webpage, instructions, public presentations, and blog posts, as well as social media posts on Facebook and Twitter. It also included provisional scientific results in the form of project updates, a process that deviated from the conventional chain of genres communicating scientific results. Such chains are typically seen as

Table 2. Digital tools used to develop and communicate the Heartbeats Project

Data analysis	Composition & communication
Data Thief (data extraction)	Google Docs (drafting)
WuFoo (online forms)	MS Word (drafting)
PanTHERIA (database)	Adobe Illustrator (figures)
AnAge (database)	Keynote (Clay's presentations)
Google Scholar (database)	MS PowerPoint (Jada's presentations/posters)
Web of Science (database)	Email (internal)
PubMed (database)	Gmail Chat (internal)
Interlibrary loan portal (article access)	Skype (internal)
Library websites (present & former institutions)	Google Hangouts (internal)
Wikipedia (for scientific names)	Lab listserv (internal)
Dropbox (data storage)	WordPress (external)
MS Excel (data storage & manipulation)	SciStarter (citizen science database)
Notepad & CSV (data storage & manipulation)	Lab and project websites (external)
Graphpad Prism (preliminary analyses)	Twitter (external)
R & RStudio (statistical analysis)	Facebook (external)
D3.js (interactive data visualization)	Google+ (external)

beginning with publication of scientific research articles some of which progress to popularizations for lay audiences as press releases and science news articles (see Fahnestock 1986; Myers 2003) or are "commodified" in science textbooks (Myers 1992). The Heartbeats Project team upended this sequence, popularizing scientific activity and results throughout their research process in order to communicate with citizen scientists. However, this communication became part of that scientific activity, rather than separate from or parallel to it, serving an inventional role in their scientific composing (Reid 2017). This transparency about their science was atypical enough that Clay had concerns that it might impact their ability to publish in a scientific journal. He explained, for example, how major journals often prohibit scientists from blogging about or presenting results (e.g. at conferences) ahead of publication. The team's engagement with citizen scientists throughout the process – engagement that included communicating results – "pushed the line" on these sorts of norms, a risk Soren and members of his lab found worthwhile in order to provide the public with more meaningful access to science.

The desire to communicate with citizen scientists about the Heartbeats Project provided Clay with a recurring exigence to deepen his analysis and return to his manuscript. During the study, for example, Clay spoke about the project at two science cafés, events that feature scientists presenting their work in casual settings such as pubs. These were opportunities to recruit new public contributors to the project as well as "influence the writing" of his manuscript. Preparing for his first science café talk pushed Clay and his coauthors to examine heartrate-longevity trends for different orders of mammals, such as marsupials and rodents. Afterwards, Clay felt that now "we more or less have what we want to put together for our paper." Applying "modern phylogenetic thinking about what's going on" for the science café had helped Clay to "organize my thoughts about where to start" and "provide the context." Although not digital moments, these cafés were enabled by a larger ecosystem of genres, media, and platforms that relied heavily on the Internet and digital media, and underscores the routine polycontextuality at work in the lab. The intentional and unintentional meshing of social contexts and genres provides a clear example of "context collusions," or consciously enabled context collapse (Davis and Jurgenson 2014). Understanding the team's scientific genre performances requires understanding this distinctly digital dynamic and the new relationships between traditional genres like scientific research articles and new genres like online citizen science instructions that characterize it.

4.2 Composing for recomposition

In addition to the team's routine use of polycontextuality for scientific invention, this dynamic was evident in the team's performance of the research article genre itself, particularly in their compositional choices around its visual elements. Not only did the team harness ubiquitous "context collapse" (Wesch 2009; Marwick and Boyd 2011) for inventional purposes, but also composed for it in their scientific genres, expecting their research articles, and specifically the figures from them, to circulate and act in multiple contexts. With this in mind, Soren and his lab designed their figures with the specificity necessary for making scientific arguments to specialized audiences, but also with elements that would make them accessible and inviting to public audiences, with the thought that the figures would be recontextualized in other texts and genres. Soren found the figures so valuable that he was "actually willing to spend a fair amount of money making a figure like this really work," a priority evident in the lab's staffing, which included a designer and a curator of digital media.

While remaining faithful to the importance of figures in scientific discourse, the lab's approach deliberately designed them for polycontextuality through the process of "composing for recomposition" (Ridolfo and DeVoss 2009). This process occurs when "composers anticipate and strategize future third-party remixing of their compositions as part of a larger and complex rhetorical strategy that plays out across physical and digital spaces" (intro.html). In addition to its focus on citizen science, Soren's lab further embraced participatory culture in their scientific composing, anticipating the repurposing of their manuscripts by diverse composers for varied audiences and in different genres. Some of these authors could include members of the lab, in a "disintermediated" approach to communicating with the public about their project, but could also include the traditional mediators of scientific knowledge for public audiences, such as their university's public information officers in the form of press releases and science journalists in the form of popular science articles that could be shared on social media.

In reference to the central figure in the Heartbeats Project manuscript, Clay explained that "if you know it's going to be something you'll use in presentations" and you know "people are going to be talking about the data in popular publications," then "why not have something that's already amenable to those formats?" Anticipating how journalists might recontextualize the manuscript, Clay mused that they might "just write about the paper and draw a picture of a heart," an illustration he felt did little to communicate their science. Instead, he hoped that by creating and publishing a figure that would communicate well with both scientists and members of the public, the team could influence how their scientific story would be told (such as "show[ing] a picture of the actual data that makes sense to people"). The Heartbeats Project team's approach to working with science journalists took advantage of the affordances of networked digital media, such as easy transmission of image files, to invoke and cultivate a scientifically literate public.

Composing figures for recomposition across contexts, in fact, was the aspect of their scientific genre performance that Soren found most significant and that he saw as most distinctively tied to new media. He explained that these figures were often simple and not obviously tied to "new technology," but that a figure designed this way incorporated many "design ideas to make it really, easily compelling." Rather than sharing data, hyperlinking to references, or including video, Soren focused on the affordances of digital media to connect with people, explaining that "the thing that *does* happen [when designing a figure for recomposition], is that a figure like this gets set up to be *part* of new media." Features of the figures that enhanced their ability to be shared across social media platforms or integrated in news stories were "not an accident," but were built with a "consciousness of this [sharing] as a discussion."

Rather than viewing his scientific research articles as solely operating within specialist discourse communities, Soren and his lab members acknowledged and wrote for their role in non-specialist communities, seeking to amplify their role by enabling public engagement and sharing with elements of the genre.

Writing research articles to facilitate sharing by non-specialist communities, however, can have unintended or undesirable consequences. Soren was aware of this possibility, even while he was generally hopeful about communicating with the public and making scientific work and conversations accessible. Ridolfo and

DeVoss (2009) address the fact that writers who compose for recomposition often "balance the future possibilities in terms of possible positive, negative, and neutral outcomes for recomposing, remixing, and appropriation" (velocity.html), weighing and trying to account for these possibilities in their writing. While he anticipated many positive outcomes for creating research article figures that would be easily shared, Soren's foremost concern was that the science might be mischaracterized or misunderstood. He was keenly aware that "the way that people are now writing about [science] stories" tended to be "under very quick timelines" and "often driven by an image" with the stories aimed "below scientific background." To avoid misunderstandings, Soren and his team attempted to create a figure that reflected "the whole narrative of the paper" and that "structures the narrative really strongly," continually asking, "How do we make the study *unbreakable* through the figure?" This question was not one he would have asked while addressing solely specialist scientific readers, but became important with the recognition of the permeability of discourse communities and the multiple purposes the same research article or figure could simultaneously play across multiple contexts and used in different genres. By composing for recomposition, Soren not only recognized that uptake of a research article might simultaneously occur in specialist and public genres, but also sought to influence that uptake.

4.3 Composing for and with the public to change science

The final dimension of the team's digital genre performances relevant to the focus of this collection relates to their complex conceptualization of communicating with the public. Rather than positioning the public solely as laypeople, the team acknowledged that fellow scientists were also members of the public and would be reached by their communication with citizen scientists. For example, Clay acknowledged that in addition to "amateur-experts" who might contribute data to the Heartbeats Project, they were likely to also receive contributions from "scientists who are just bored at work," particularly since they were more likely to have access to the journal articles necessary to contribute. Clay noted that he and his colleagues were constantly attuned to the fact that "scientists are all just people" with the same human interests as other people. Based on the questions he had received, it was clear that some members of the audience at Clay's first science café talk were scientists, including a "cranky" one who had stayed afterwards to question Clay's methodology.

Beyond acknowledging the routine polycontextuality at work in communicating in public fora, however, the team had actually counted on it, using this fact for a variety of goals. One of these was simply to ensure that fellow scientists learned about their work, ideally using it themselves. For example, Brendan, the university public information officer the team worked with, explained that several studies supported the notion that scientific popularizations reached scientists, increasing the chances of scientists' work impacting future science (see, for instance, Phillips et al. 1991; Kiernan 2003; Liang et al. 2014). This phenomenon had been documented prior to widespread access to the Internet and digital technologies, but was amplified by the contemporary media environment, with news media, social media, scientist blogs, and scientific publications operating synergistically.

A more powerful motivation for Soren, however, was influencing science itself (especially what he saw as a limiting hyperspecialization and "atomization") by composing for and with the public. Soren regretted the hyperspecialization of science, explaining that "the ways of being a scientist have atomized," with many scientists not even noticing the phenomenon because "when they're being trained," they "see this one domain and the adjacent domains," but no further. This atomized specialization was sustained by a variety of professional structures and incentives, something that Soren stated could be "productive," but not necessarily satisfying or conducive to some valuable types of scientific thinking and research.

For Soren, citizen science constructed and performed a science defined by common interests and access over division and specialization. As Soren put it, using digital media to reach multiple audiences at once – all invoked in the role of "citizen scientist" "lends itself to new ways of being part of a broader community in a moment in which science has become, maybe not as specialized as we think it is, but pretty specialized." Beyond a public context, citizen science functioned to purposefully collapse multiple scientific contexts and genres – intraspecialistic, interspecialistic, pedagogical, and popular (Shinn and Cloître 1985) and illustrates the "pragmatic openness" of the Internet (Giltrow and Stein 2009: 9) – and its potential for "creat[ing] new functions" (10). In what he saw as a "hopeful" move, Soren wanted to reach scientists in other disciplines and to demonstrate both to them and to his postdocs and students that "we can still bridge these [divisions], but to do it, we're going to use different kinds of tools" and have to learn to "talk to other people in their own language." A clear example of "context collusion," Soren's citizen science communication made an implicit argument for rendering such discursive meshing routine and recurrent, and for expanding the possible habitus of an effective scientist to include this, along with the broader inquiry such meshing enables. In his lab manager's words, Soren modeled a "digital da Vinci."

5. Conclusion

Recent advances in writing studies have explored what Prior and Shipka (2003: 181) call the "chronotopic laminations" of people's literate activity – the "dispersed, fluid chains of places, times, people, and artifacts that come to be tied together in trajectories of literate action along with the ways multiple activity footings are held and managed." Methodologically, Prior and Shipka (2003: 231) note, "private and public acts, meaning and sense, affect and attention, tools and spaces, all need to be woven together into a single story of productive activity." Writing research grounded in these assumptions of activity theory has taken the form of case studies that examine the ways that multiple contexts, genres, audiences, and preoccupations or interests become enmeshed in the various texts that individuals write, sometimes tacitly and sometimes consciously influencing their work and/or becoming invisible to readers (see, for example, Roozen 2010). The results demonstrate that writers do not work in rhetorical or even physical spaces and genres sealed off from the diversity of psychological, social, experiential/historical, and contextual factors that play a role in their texts.

At the same time, and as a way to demonstrate the permeability of these spaces, much of this research has examined the ways that two domains interact or interrelate – for example, how college students' academic writing is influenced by their nonacademic experiences, such as their participation in online communities or other personal interests that inform their composing. The research we report here extensively examined what prior scholarship would have considered a relatively "closed" context – specialized scientific writing produced in a scientific laboratory – but in light of the Heartbeat Project's deliberate attempts to reciprocally inform and be informed by public audiences. The results, however, support a more complex model of genre that moves beyond the influence of one domain on another (e.g. nonacademic on academic, or scientific on public) and the common representation of these influences as tacit. Instead, the model admits a willing multidirectionality of influence: the team (1) composed public genres as part of the activity of doing (specialized) science (public → science); (2) composed scientific genres with an eye toward facilitating public discourse and the ability of specialized science to act in public contexts (science → public); and (3) used public-science collaboration in the form of citizen science genres to shape the scientific community (public ←→ science), with the hopes of redefining it to be more capacious and integrated, and to therefore impact the possible ways of "being a scientist." Single acts of composing were intended to act in multiple genre ecosystems (polycontextuality) and achieve multiple functions at once, with Soren and some members of his team aware of the constitutive dimension of their communication: they hoped to constitute a less atomized scientific community with room for various levels of participation, including that of citizen amateur-experts and novices. In this sense, Soren was hyper-aware of the permeability between genres and purposefully emphasizing and constructing that permeability through his science communication. For him, "context collapse" was a principal affordance of digital media for science, and he actively "colluded" to accentuate it because of his complex (democratic) view of science and publics. In his view, the public was integrally bound up with science, and demarcations between scientific specialties and between science and public were artificial and potentially harmful to both.

The case illustrates sophisticated, creative strategies for effectively navigating the digital age as a scientist, but also illustrates sophisticated awareness of the constitutive dimension of communication – Soren was not just "navigating" the current media environment, but also aware of shaping it through his communication. He was actively shaping a different kind of scientific community and different "ways of being" scientists, aware that uptake of his genre performances and genre repertoire could encourage recurrence and eventually typify different possible roles for scientists and interested citizens. Keenly aware of how integrally bound up science and the public are with each other's past, present, and future, he included public digital genres in his repertoire of scientific communication as they enabled the (re)production of science and of a certain variety of scientific inquiry that he found valuable.

Finally, we acknowledge the idiosyncrasies of all scientific contexts, populated as they are by unique individuals who bring diverse dispositions, skills, and orientations to their work; assisted as they are (or not) by the affordances of specific communication technologies; and motivated as they are (or not) by ideologies of the relationships between scientific inquiry and the public interest. Soren's agentive "context collusion" serves as a reminder of the role of individuals in genre stability and change and against the impulse to systematize too strongly the relationships between genres. Further study of unique groups of scientists and the publics with which they interact, with attention to their respective genre performances as part of that interaction, will continue to strengthen our current understanding of scientific communication, while it can be hoped that those interactions lead to a more informed and participatory citizenry and a scientific community that values its contributions to that effort.

References

Anson, Chris M., and Deanna Dannels. 2004. "Writing and Speaking in Conditional Rhetorical Space." In *Classroom Space(s) and Writing Instruction*, ed. by Ed Nagelhout, and Carol Rutz, 55–70. Cresskill, NJ: Hampton Press.

Bazerman, Charles. 1997. "Discursively Structured Activities." *Mind, Culture, and Activity* 4: 293–308. https://doi.org/10.1207/s15327884mca0404_6

Beaulieu, Anne. 2010. "Research Note: From Co-Location to Co-Presence: Shifts in the Use of Ethnography for the Study of Knowledge." *Social Studies of Science* 40 (3): 453–470. https://doi.org/10.1177/0306312709359219

Beaufort, Anne. 2007. *College Writing and Beyond: A New Framework for University Writing Instruction*. Logan: Utah State University Press.

Bizzell, Patricia. 2009. *Academic Discourse and Critical Consciousness*. Pittsburgh: University of Pittsburgh Press.

Bruckman, Amy. 2002. "Studying the Amateur Artist: A Perspective on Disguising Data Collected in Human Subjects Research on the Internet." *Ethics and Information Technology* 4 (3): 217–231. https://doi.org/10.1023/A:1021316409277

Charmaz, Kathy. 2003. "Qualitative Interviewing and Grounded Theory Analysis." In *Inside Interviewing: New Lenses, New Concerns*, ed. by James A. Holstein, and Jaber F. Gubrium, 311–30. Thousand Oaks, CA: Sage.

Cummings, Jonathon N., and Sara Kiesler. 2005. "Collaborative Research across Disciplinary and Organizational Boundaries." *Social Studies of Science* 35 (5): 703–722. https://doi.org/10.1177/0306312705055535

Davies, Sarah R., and Noriko Hara. 2017. "Public Science in a Wired World: How Online Media Are Shaping Science Communication." *Science Communication* 39 (5): 563–68. https://doi.org/10.1177/1075547017736892

Davis, Jenny L., and Nathan Jurgenson. 2014. "Context Collapse: Theorizing Context Collusions and Collisions." *Information, Communication & Society* 17 (4): 476–85. https://doi.org/10.1080/1369118X.2014.888458

Donahue, Tiane. 2017. "Writing and Global Transfer Narratives: Situating the Knowledge Transformation Conversation." In *Critical Transitions: Writing and the Question of Transfer*, ed. by Chris M. Anson, and Jessie L. Moore, 107–136. Anderson, SC: Parlor Press and the WAC Clearinghouse.

Engeström, Yrjö, Ritva Engeström, and Merja Kärkkäinen. 1995. "Polycontextuality and Boundary Crossing in Expert Cognition: Learning and Problem Solving in Complex Work Activities." *Learning and Instruction* 5 (4): 319–336. https://doi.org/10.1016/0959-4752(95)00021-6

Fahnestock, Jeanne. 1986. "Accommodating Science: The Rhetorical Life of Scientific Facts." *Written Communication* 3 (3): 275–296. https://doi.org/10.1177/0741088386003003001

Giltrow, Janet, and Dieter Stein (eds.). 2009. *Genres in the Internet: Issues in the Theory of Genre*. Philadelphia: John Benjamins. https://doi.org/10.1075/pbns.188

Glaser, Barney G., and Anselm L. Strauss. 1967. *The Discovery of Grounded Theory: Strategies for Qualitative Research*. New York: Aldine de Gruyter.

Gupta, Akhil, and James Ferguson. 1997. "Discipline and Practice: 'The Field' as Site, Method, and Location in Anthropology." In *Anthropological Locations: Boundaries and Grounds of a Field Science*, ed. by Akhil Gupta, and James Ferguson, 1–47. Berkeley and Los Angeles: University of California Press.

Hine, Christine. 2006. "Databases as Scientific Instruments and Their Role in the Ordering of Scientific Work." *Social Studies of Science* 36 (2): 269–98. https://doi.org/10.1177/0306312706054047

Hine, Christine. 2015. *Ethnography for the Internet: Embedded, Embodied and Everyday*. London: Bloomsbury.

Kiernan, Vincent. 2003. "Diffusion of News about Research." *Science Communication* 25 (1): 3–13. https://doi.org/10.1177/1075547003255297

Levine, Herbert J. 1997. "Rest Heart Rate and Life Expectancy." *Journal of the American College of Cardiology* 30 (4): 1104–1106.

Lauer, Janice M. 2004. *Invention in Rhetoric and Composition*. West Lafayette, IN: Parlor Press.

Liang, Xuan, Leona Yi-Fan Su, Sara K. Yeo, Dietram A. Scheufele, Dominique Brossard, Michael Xenos, Paul Nealey, and Elizabeth A. Corley. 2014. "Building Buzz: (Scientists) Communicating Science in New Media Environments." *Journalism & Mass Communication Quarterly* 91 (4): 772–791. https://doi.org/10.1177/1077699014550092

Lillis, Theresa. 2008. "Ethnography as Method, Methodology, and 'Deep Theorizing': Closing the Gap between Text and Context in Academic Writing Research." *Written Communication* 25 (3): 353–388. https://doi.org/10.1177/0741088308319229

Lillis, Theresa, and Janet Maybin. 2017. "Introduction: The Dynamics of Textual Trajectories in Professional and Workplace Practice." *Text & Talk* 37 (4): 409–414. https://doi.org/10.1515/text-2017-0017

Marincola, Elizabeth. 2006. "Why is Public Science Education Important?" *Journal of Translational Medicine* 4: 7. https://doi.org/10.1186/1479-5876-4-7

Marwick, Alice E., and Danah Boyd. 2011. "I Tweet Honestly, I Tweet Passionately: Twitter Users, Context Collapse, and the Imagined Audience." *New Media & Society* 13 (1): 114–33. https://doi.org/10.1177/1461444810365313

Myers, Greg. 1990. *Writing Biology: Texts in the Social Construction of Scientific Knowledge.* Madison: University of Wisconsin Press.

Myers, Greg. 1992. "Textbooks and the Sociology of Scientific Knowledge." *English for Specific Purposes*, 11, 3–17. https://doi.org/10.1016/0889-4906(92)90003-S

Myers, Greg. 2003. "Discourse Studies of Scientific Popularization: Questioning the Boundaries." *Discourse Studies* 5 (2): 265–79. https://doi.org/10.1177/1461445603005002006

Odell, Lee, Dixie Goswami, and Anne Herrington. 1983. "The Discourse-Based Interview: A Procedure for Exploring the Tacit Knowledge of Writers in Nonacademic Settings." In *Research on Writing: Principles and Methods*, ed. by Peter Mosenthal, Lynne Tamor, and Sean A. Walmsley, 221–36. New York: Longman.

Pham, Daniel. 2016. "Public Engagement is Key for the Future of Science Research." *npj Science of Learning* 1, Article no: 16010. https://doi.org/10.1038/npjscilearn.2016.10

Phillips, David P., Elliot J. Kanter, Bridget Bednarczyk, and Patricia L. Tastad. 1991. "Importance of the Lay Press in the Transmission of Medical Knowledge to the Scientific Community." *New England Journal of Medicine* 325 (16): 1180–1183. https://doi.org/10.1056/NEJM199110173251620

Porter, James E. 1986. "Intertextuality and the Discourse Community." *Rhetoric Review* 5 (1): 34–47. https://doi.org/10.1080/07350198609359131

Prior, Paul, and Jody Shipka. 2003. "Chronotopic Lamination: Tracing the Contours of Literate Activity." In *Writing Selves/Writing Societies: Research From Activity Perspectives*, ed. by Charles Bazerman and David R. Russell, 180–238. Fort Collins, CO: WAC Clearinghouse.

Rainie, Lee, Cary Funk, and Monica Anderson. 2015. "How Scientists Engage the Public." Pew Research Center, Internet and Technology. Last accessed June 15, 2019. http://www.pewinternet.org/2015/02/15/how-scientists-engage-public/

Reid, Gwendolynne. 2017. "Shifting Networks of Science: Citizen Science and Scientific Genre Change." In *Scientific Communication: Practices, Theories, and Pedagogies*, ed. by Han Yu, and Kathryn Northcut. New York: Routledge.

Ridolfo, Jim, and Dànielle N. DeVoss. 2009. "Composing for Recomposition: Rhetorical Velocity and Delivery." *Kairos* 13 (2): n.p.

Roozen, Kevin. 2010. "Tracing Trajectories of Practice: Repurposing in One Student's Developing Disciplinary Writing Processes." *Written Communication* 27 (3): 318–354. https://doi.org/10.1177/0741088310373529

Russell, David R. 1995. "Activity Theory and Its Implications for Writing Instruction." In *Reconceiving Writing, Rethinking Writing Instruction*, ed. by Joseph Petraglia, 51–77. Hillsdale, NJ: Erlbaum.

Saldaña, Johnny. 2009. *The Coding Manual for Qualitative Researchers*. London: Sage.

Shinn, Terry, and Michel Cloître. 1985. "Expository Practice: Social, Cognitive and Epistemological Linkage." In *Expository Science: Forms and Functions of Popularization*, ed. by Terry Shinn, and Richard P. Whitley, 31–60. Dordrecht, Netherlands: D. Reidel. https://doi.org/10.1007/978-94-009-5239-3

Stilgoe, Jack, Simon J. Lock, and James Wilsdon. 2014. "Why Should We Promote Public Engagement With Science?" *Public Understanding of Science* 23 (1): 4–15. https://doi.org/10.1177/0963662513518154

Swales, John M. 2004. *Research Genres: Explorations and Applications*. New York: Cambridge University Press. https://doi.org/10.1017/CBO9781139524827

Trench, Brian. 2008. "Internet: Turning Science Communication inside-out?" In *Handbook of Public Communication of Science and Technology*, ed. by Massimiano Bucchi, and Brian Trench, 185–98. New York: Routledge.

Wesch, Michael. 2009. "YouTube and You: Experiences of Self-Awareness in the Context Collapse of the Recording Webcam." *Explorations in Media Ecology* 8 (2): 19–34.

Zachry, Mark. 2000. "Communicative Practices in the Workplace: A Historical Examination of Genre Development." *Journal of Technical Writing and Communication* 30 (1): 57–79. https://doi.org/10.2190/UMGD-LGR6-QJUE-CJHY

Index